Mathematics for Engineers
(Part 3)

Ram Bilas Misra
and
Sadanand Pandey

CWP

Central West Publishing

Other books of interest

Mathematics for Engineers and Physicists, Part 1
Ram Bilas Misra
ISBN (print): 978-1-925823-51-6
ISBN (e-book): 978-1-925823-50-9

Mathematics for Engineers and Physicists, Part 2
Ram Bilas Misra
ISBN (print): 978-1-925823-53-0
ISBN (e-book): 978-1-925823-52-3

Mathematics for Engineers and Physicists (Part 3)

Prof. Dr. Ram Bilas Misra

Ex Vice Chancellor, Avadh University, Faizabad, U.P. (India);
Professor of Mathematics, Lebanese French University, Erbil, Kurdistan (Iraq).

Former: *Dean*, Faculty of Science, A.P. Singh University, Rewa, M.P. (**India**);
Prof., Dept. of Maths., Higher College of Edn., Aden Univ., Aden (**Yemen**);
Professor & Head, Dept. of Maths. & Stats., A.P.S. University, Rewa, M.P. (**India**);
Prof., Dept. of Maths., College of Science, Salahaddin University, Erbil (**Iraq**);
UGC Visiting Prof., Mahatma Gandhi Kashi *Vidyapith*,Varanasi, U.P. (**India**);
Professor, Dept. of Maths, Ahmadu Bello Univ., Zaria (**Nigeria**) – designate;
Prof. & Head, Dept. of Maths. & Comp. Sci., Univ. of Asmara, Asmara (**Eritrea**);
Director, Unique Inst. of Business & Technol., Modi Nagar, Ghaziabad, U.P. (**India**);
Prof. & Head, Dept. of Maths., Phys. & Stats., Univ. of Guyana, Georgetown (**Guyana**);
Prof. & Head, Dept. of Maths., Eritrea Inst. of Technology, Mai Nefhi (**Eritrea**);
Prof.& Head, Dept. of Maths., School of Engg., Amity Univ., Lucknow, U.P. (**India**);
Prof. & Head, Dept. of Maths. & Comp. Sci., PNG Univ. of Technology, Lae (**PNG**);
Prof. of Maths., Teerthankar Mahaveer University, Moradabad, U.P. (**India**);
Prof., Dept. of Maths, Oduduwa Univ., Ipetumodu, Osun State (**Nigeria**) – designate;
Prof., Dept. of Maths, Adama Science & Technology Univ., Adama (**Ethiopia**);
Prof. & Head, Dept. of Maths. & C.S., Bougainville Inst. of Bus. & Tech., Buka (**PNG**) – designate;
Prof. & Head, Dept. of Maths., J.J.T. University, Jhunjhunu, Rajasthan (**India**);
Dean, Faculty of Science, J.J.T. University, Jhunjhunu, Rajasthan (**India**);
Professor, Dept. of Maths, Wollo University, Dessie, Wollo (**Ethiopia**);
Professor, Dept. of Appld. Maths., State Univ. of New York, Incheon (**S. Korea**)
Prof., Dept. of Maths. & Computing Sci., Divine Word Univ., Madang (**PNG**);
Director, Maths., School of Sci. & Engg., Univ. of Kurdistan Hewler, Erbil (**Iraq**);
DAAD Fellow, University of Bonn, Bonn (**Germany**);
Visiting Professor, University of Turin, Turin (**Italy**);
Visiting Professor, University of Trieste, Trieste (**Italy**);
Visiting Professor, University of Padua, Padua (**Italy**);
Visiting Professor, International Centre for Theoretical Physics, Trieste (**Italy**);
Visiting Professor, University of Wroclaw, Wroclaw (**Poland**);
Visiting Professor, University of Sopron, Sopron (**Hungary**);
Reader, Dept. of Maths. & Stats., South Gujarat University, Surat, Gujarat (**India**);
Reader, Dept. of Maths. & Stats., University of Allahabad, Allahabad, U.P. (**India**);
Asst. Prof., Dept. of Maths., College of Sci., Mosul Univ., Mosul (**Iraq**) – designate;
Senior most *NCC Officer* (Naval Wing), Univ. of Allahabad, Allahabad, U.P. (**India**);
Lecturer, Dept. of Maths., KKV Degree College, Lucknow, U.P. (**India**).

Prof. Dr. Sadanand Pandey

Former
Prof. of Applied Mathematics & Head, Dept. of Humanities,
M.M.M. Engg. College (now University), Gorakhpur, U.P. (**India**);
Prof., Dept. of Maths. & Comp. Sci., Univ. of Asmara, Asmara (**Eritrea**);
Prof. & Head, Dept. of Maths. & Comp. Sci., PNG Univ. of Tech., Lae (**PNG**);
Visiting Professor, International Centre for Theoretical Physics, Trieste (**Italy**).

A catalogue record for this book is available from the National Library of Australia

NATIONAL LIBRARY OF AUSTRALIA

ISBN (print): 978-1-925823-64-6
ISBN (ebook): 978-1-925823-66-0

CONTENTS

PREFACE

The book covers the topics of Mathematics syllabus of the third semester of bachelor's degree in main engineering courses in most of the universities all over the world. Knowledge of foundation courses in basic college mathematics such as classical algebra, trigonometry, 2-dimensional coordinate geometry, linear algebra, calculus (both differential and integral) of real functions, ordinary differential equations (ODEs), Beta and Gamma functions, and elementary statistics are a pre-requisite. For brevity, some set-theoretic notions and symbols are frequently used, e.g. the symbol $\Rightarrow$ means *implies*. The *logarithm* of a number to the exponential base e is denoted by ln. All the Latin mathematical symbols are normally *italicized*, while their Greek counterparts are in normal fonts. A selected bibliography of the subject is provided. The alphabetical index added at the end makes access to the contents easier.

The subject matter is presented here in *ten* chapters of which the first one includes the results used in the later discussion. The basis of Chapters 2-8 originates from my earlier texts, while most of the contents in the last two chapters are originally drafted by the second author. Contents of the chapters are divided into Sections and the discussion within the Sections is presented in the form of Definitions, Theorems, Corollaries, Notes and Examples. Most of the chapters end in a Problem Set containing unsolved exercises with solutions to challenging ones. The sub-titles within the Sections are numbered in decimal pattern. For instance, the equation number $(c.s.e)$ refers to the eth equation in the sth section of Chapter c. When c coincides with the chapter at hand, it is dropped. Adequate references to the results appeared earlier are made in the text avoiding their unnecessary repetition. Double slashes marked at the end of Theorems, Corollaries, Solutions of Exercises, etc., indicate their completion.

Author's long teaching career of more than ***five decades*** at various universities round the globe and research expertise in different fields helped him for lucid presentation of the subject. In the preparation of the text, I have immensely benefited through books at sr. nos. [11-19] in the bibliography.

The book is dedicated to teachers, mentors, colleagues and friends of the first author who helped him in multifold life. Due to the good care by kith & kin, even the advanced age could not prevent me to set aside my passion for the subject, hence, they also deserve my sincere appreciation. Thanks are also due to various universities all over the world especially University of Allahabad, Prayagraj (Allahabad, India); University of Asmara, Asmara (Eritrea), University of Guyana, Georgetown (Guyana); P.N.G. University of Technology, Lae (Papua New Guinea); Adama Science & Technology University, Adama (Ethiopia); State University of New York, Incheon (South Korea); Divine Word University, Madang (P.N.G.), and my present institution (LFU), where I gained a lot in exposing my expertise. Sincere thanks are also due to the publisher for their valuable cooperation for bringing the book into limelight in a limited time.

Although proofs are read with utmost care and solutions to problems are verified repeatedly, yet an oversight or any discrepancy brought to the notice of the author by the inquisitive readers(s) shall be thankfully acknowledged. What a surprising coincidence of completing the first draft of the manuscript on the 4th day of most auspicious period of *Navratri* in conformity of the belief of my parents for which the Mother Goddess Durga has helped me to accomplish it without even getting slightest indisposition.

Lucknow (India): May 10, 2019   Ram Bilas Misra

CHAPTER 1

PRELIMINARIES

§ 1. Results referred in the text

The results and formulae used in the text are stated below without proof. The proof of such results may be found in author's previous texts referred to in the Bibliography, cf. [11] – [18].

1.1. Calculus (Differential)

Theorem 1.1. Given a function $f(x)$ defined in a closed interval $[a, b]$ on real line, there exists at least one zero of $f(x)$, say at point x_1, in the interval if $f(a)$ and $f(b)$ have opposite signs.

Theorem 1.2. (Rolle's theorem) If a function $f(x)$ is continuous in a closed interval $[a, b]$, differentiable in the open interval (a, b), and satisfying $f(a) = f(b) = 0$, then there exists at least one point, say x_1, in the open interval (a, b), where $f'(x_1) = 0$.

Theorem 1.3. (Intermediate value theorem) Given a continuous function $f(x)$ defined in a closed interval $[a, b]$, there exists a point, say x_1, in the open interval (a, b), where holds $a < f(x_1) < b$.

Theorem 1.4. (Mean value theorem) If a function $f(x)$ is continuous in a closed interval $[a, b]$, differentiable in the open interval (a, b), then there exists at least one point, say x_1, in the open interval (a, b), where

$$f'(x_1) = \{f(b) - f(a)\} / (b - a). \tag{1.1}$$

Taking $b = a + h$, above result also reads as

$$f(a + h) = f(a) + h f'(a + h\theta), \ 0 < \theta < 1. \tag{1.2}$$

Theorem 1.5. (Maclaurin's theorem) Given a continuous function $f(x)$ possessing continuous derivatives of any order at the origin, then it is expandable in powers of x :

$$f(x) = f(0) + x . f'(0) + (x^2/2!) . f''(0) + \ldots + (x^n/n!) . f^{(n)}(0) + \ldots . \tag{1.3}$$

Theorem 1.6. (Taylor's theorem for a function of one variable) Given a continuous function $f(x)$ possessing continuous derivatives of any order at a given point, say $x = a$, then it is expandable in powers of $x - a$:

$$f(x) = f(a) + (x - a).f'(a) + \{ (x - a)^2/2! \}.f''(a) + \dots$$

$$+ \{ (x - a)^n/n! \}.f^{(n)}(a) + \dots ; \tag{1.4}$$

alternately, for $x = a + h$,

$$f(a + h) = f(a) + h.f'(a) + (h^2/2!).f''(a) + \dots$$

$$+ \{ (h)^n/n! \}.f^{(n)}(a) + \dots ; \tag{1.4a}$$

also, replacing a by x

$$f(x + h) = f(x) + h.f'(x) + (h^2/2!).f''(x) + \dots$$

$$+ \{ (h)^n/n! \}.f^{(n)}(x) + \dots . \tag{1.4b}$$

Theorem 1.7. (Taylor's theorem for a function of two variables) Given a continuous function $f(x, y)$ of two variables possessing its continuous partial derivatives of any order at a given point, say P (x, y), then as a generalization of Eq. (1.4b), there holds:

$$f(x + h, y + k) = f(x, y) + \{ h.(\partial f/\partial x) + k.(\partial f/\partial y) \}$$

$$+ (1/2!). \{ h.(\partial/\partial x) + k.(\partial/\partial y) \}^2 f + \dots$$

$$+ (1/n!). \{ h.(\partial/\partial x) + k.(\partial/\partial y) \}^n f + \dots . \tag{1.5}$$

Theorem 1.8. (Taylor's theorem for a function of n variables) Given a continuous function $f(x_1, x_1, \dots, x_n)$ of n variables possessing its continuous partial derivatives of any order there holds generalization of Eq. (1.5):

$$f(x_1 + \Delta x_1, x_2 + \Delta x_2, \dots, x_n + \Delta x_n) = f(x_1, x_1, \dots, x_n)$$

$$+ \{ \Delta x_1 (\partial/\partial x_1) + \Delta x_2 (\partial/\partial x_2) + \dots + \Delta x_n (\partial/\partial x_n) \} f$$

$$+ (1/2!). \{ \Delta x_1 (\partial/\partial x_1) + \Delta x_2 (\partial/\partial x_2) + \dots + \Delta x_n (\partial/\partial x_n) \}^2 f + \dots$$

$$+ (1/n!).\{ \Delta x_1 (\partial/\partial x_1) + \Delta x_2 (\partial/\partial x_2) + \dots + \Delta x_n (\partial/\partial x_n) \}^n f + \dots \tag{1.6}$$

1.2. Calculus (Integral)

There hold the following properties of definite integrals:

$$\int_{-a}^{a} f(x)dx = 2 \int_{0}^{a} f(x)dx, \quad \text{or} \quad 0, \tag{1.7}$$

when $f(x)$ is an even (or odd) function of x, i.e. $f(-x) = \pm f(x)$;

$$\int_{x=0}^{2a} f(x)dx = 2 \int_{x=0}^{a} f(x)dx, \quad \text{or} \quad 0, \tag{1.8}$$

when $f(2a - x) = f(x)$ or $-f(x)$ respectively.

Theorem 1.9. We have

$$\int (e^{ax} \sin bx) \, dx = (a^2 + b^2)^{-1} e^{ax} (a \sin bx - b \cos bx), \tag{1.9}$$

and

$$\int (e^{ax} \cos bx) \, dx = (a^2 + b^2)^{-1} e^{ax} (a \cos bx + b \sin bx). \tag{1.10}$$

1.3. Coordinate geometry (2-dimensional)

A straight line passing through two points P_1, P_2 having rectangular Cartesian coordiantes (x_1, y_1) and (x_2, y_2) respectively has equation

$$y - y_1 = \{(y_2 - y_1) / (x_2 - x_1)\} . (x - x_1). \tag{1.11}$$

1.4. Beta and Gamma functions

Leonid (Leonhard in German) Euler (15.4.1707 – 18.9.1783) gave two kinds of integrals called *Beta* and *Gamma functions*: the first kind integral, denoted by $B(m, n)$, is called the Beta function:

$$B(m, n) \equiv \int_{0}^{1} x^{m-1} . (1-x)^{n-1} \, dx, \quad m > 0, \ n > 0; \tag{1.12}$$

and the second kind, denoted by $\Gamma(m)$, is called the Gamma function:

$$\Gamma(m) \equiv \int_{0}^{\infty} e^{-x} . x^{m-1} . dx, \quad m > 0. \tag{1.13}$$

These functions are connected by

$$B(m, n) = \Gamma(m) . \Gamma(n) / \Gamma(m+n). \tag{1.14}$$

Also, Gamma function satisfies

$$\Gamma(m+1) = m.\,\Gamma(m) = m!, \text{ for positive integer } m; \qquad (1.15)$$

$$\Gamma(1) = 1, \qquad\qquad \Gamma(1/2) = \sqrt{\pi}. \qquad (1.16)$$

Note 1.1. The name Beta function was later given by Legendre, Whittaker and Watson. Daniel Bernoulli was the first person to have introduced an integral (later called as Gamma function) in a letter to Goldbach in 1729. These functions are also written with small letters b (respectively g). Historically, Euler used the Greek capital (letter) B for beta function.

CHAPTER 2

SOLUIONS OF ALGEBRAIC AND TRANSCENDENTAL EQUATIONS

§ 1. Introduction

Most complex problem in scientific and engineering studies is to seek solutions of a general equation in one variable

$$f(x) = 0. \tag{1.1}$$

In case, the function $f(x)$ is algebraic, i.e. a polynomial of some finite degree, say n, there are algebraic methods to find the solutions of above equation. On contrary, if it is transcendental or combination of algebraic and transcendental functions involving logarithmic or trigonometric functions, the roots of Eq. (1.1) may be real or complex numbers. In the following we attempt to compute roots of such equations.

1.1. Bisection method

This method is based on Theo. 1.1.1. Let $f(a)$ be negative and $f(b)$ positive. As such, the theorem assures existence of a real root, say x_0. Let its approximate value be the average of a and b:

$$x_0 = (a + b) / 2. \tag{1.2}$$

If the Eq. (1.1) gets satisfied by x_0, i.e. $f(x_0) = 0$, it is the desired root else the actual root may lie either between a and x_0 or between x_0 and b depending on if $f(x_0)$ is negative or positive respectively. Let the interval (a, x_0) alternately (x_0, b) be denoted as $[a_1, b_1]$ which is of length $|b - a|/2$. Let x_1 be the mid-point of new interval with length half of that of the previous interval. It may be the solution of Eq. (1.1), if $f(x_1) = 0$, else $f(x_1) < 0$, or > 0. The process is repeated continuously in order to get a better approximation for the desired root.

Note 1.1. A convenient method to compute the percentage error in the values of roots is given by

$$\varepsilon_r = |(x_{r+1} - x_r) / x_{r+1}| \times 100\,\%. \tag{1.3}$$

The following Examples demonstrate the process more convincingly.

Example 1.1. Compute a real root of the cubic equation

$$f(x) = x^3 + x - 1 = 0. \tag{1.4}$$

Solution. We note that the function is negative at $x = 0$, but positive at $x = 1$. So, by Theo. 1.1.1, the root of Eq. (1.4) lies in the interval $(0, 1)$. Let its approximate value be the bisecting point of the interval: $x_0 = (0 + 1) / 2 = 1/2$. Since

$$f(x_0) = x_0^3 + x_0 - 1 = 1/8 + 1/2 - 1 = -3/8 < 0,$$

$y = f(x)$

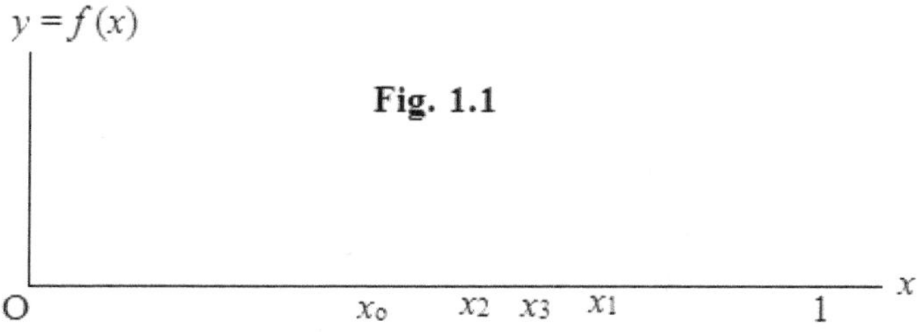

Fig. 1.1

the root lies in the interval $(1/2, 1)$. Taking the mid-point of the interval as its approximate value

$$x_1 = (1/2 + 1)/2 = 3/4, \tag{1.5}$$

where

$$f(x_1) = x_1^3 + x_1 - 1 = 27/64 + 3/4 - 1 = 11/64 > 0,$$

implies that the root lies in the interval $(1/2, 3/4)$, say at the mid-point of the interval: $x_2 = (1/2 + 3/4)/2 = 5/8$, where

$$f(x_2) = x_2^3 + x_2 - 1 = 125/512 + 5/8 - 1 = -67/512 < 0,$$

implies that the root lies in the interval $(5/8, 3/4)$, say at the mid-point of the interval: $x_3 = (5/8 + 3/4)/2 = 11/16$, where

$$f(x_3) = x_3^3 + x_3 - 1 = 1331/4096 + 11/16 - 1 = 51/4096 > 0,$$

implies that the root lies in the interval $(5/8, 11/16)$, say at the mid-point of the interval: $x_4 = (5/8 + 11/16)/2 = 21/32$, etc. //

Example 1.2. Compute the positive root of the equation

$$f(x) = x e^x - 1 = 0, \tag{1.6}$$

lying in the interval $(0, 1)$.

Solution. For $f(0) = -1 < 0$ and $f(1) = e - 1 \approx 1\cdot718 > 0$, it follows from Theo. 1.1.1, that a positive real root of Eq. (1.6) lies in the interval $(0, 1)$. Taking its approximate value at the mid-point of the interval, i.e. $x_0 = (0 + 1)/2 = 0\cdot5$. Next, at this point

$$f(x_0) = 0\cdot5. \, e^{0\cdot5} - 1 = \sqrt{e}\,/2 - 1 \approx 1\cdot648/2 - 1 = -0\cdot176 < 0,$$

implying, again by Theo. 1.1.1, that the root lies in the interval $(1/2, 1)$. Taking its approximate value at the mid-point of the interval, as in Eq. (1.5), say $x_1 = 0\cdot75 = 3/4$. Thus, the error in this (approximate) value, by Eq. (1.3), may be found:

$$\varepsilon_1 = |(x_1 - x_0)/x_1| \times 100\,\% = (0\cdot25/0\cdot75) \times 100\,\% = 33\cdot33\,\%.$$

Again, for $f(x_1) = 0\cdot75. \, e^{0\cdot75} - 1 = (3/4) \, e^{3/4} - 1 \approx 1\cdot117 > 0$, the root lies in the interval $(0\cdot50, 0\cdot75)$ and as in the previous example,

$$x_2 = (0\cdot50 + 0\cdot75)/2 = 0\cdot625.$$

The error in this root is

$$\varepsilon_2 = |(x_2 - x_1)/x_2| \times 100\,\% = (0\cdot125/0\cdot625) \times 100\,\% = 20\,\%.$$

Proceeding in the same way, we can compute better approximations to the desired root. //

§ 2. Regula-falsi method

Regula falsi (in Latin), i.e. the 'false position method' is very old method for solving a non-linear equation in one unknown. In other words, it is the trial and error technique of using test values for the variable and later adjusting these values according to the outcome.

For instance, seeking a solution of the equation $f(x) = x + x/4 = 15$, let $x = 4$ makes the solution, for which LHS $= 4 + 4/4 = 5$. Though it gives an integral value but, 4 is not the solution of the original equation, as $f(4)$ is much lesser than 15. To make it, we multiply x (currently set as 4) by 3 and substitute again to make 15. Thus, the correct solution of the equation is $x = 12$.

Modern versions of this technique employ systematic ways of choosing new test values and are concerned with the questions of whether or not an approximation to a solution can be obtained; and if it can, how fast can the approximation be found?

Theorem 2.1. An approximate solution of equation (1.1) is given by

$$x_1 = \{a_0 f(b_0) - b_0 f(a_0)\} / \{f(b_0) - f(a_0)\}, \tag{2.1}$$

where A $\{a_0, f(a_0)\}$ and B $\{b_0, f(b_0)\}$ are two points on the graph of the curve represented by the equation such that $f(a_0) < 0$ but $f(b_0) > 0$.

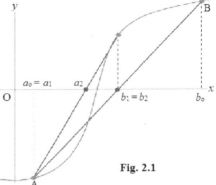

Fig. 2.1

Proof. Equation of the chord joining two points A, B is

$$\frac{x - a_0}{b_0 - a_0} = \frac{y - f(a_0)}{f(b_0) - f(a_0)}.$$

It intersects x-axis, say in a point $\{x_1, f(x_1) = 0\}$. Thus, making $y = 0$, above equation yields

$$x_1 = a_0 - (b_0 - a_0). f(a_0) / \{f(b_0) - f(a_0)\},$$

or in the form of Eq. (2.1). //

Example 2.1. Compute a real root of the cubic equation

$$f(x) \equiv x^3 - x - 3 \equiv x(x-1)(x+1) - 3 = 0. \tag{2.2}$$

Solution. We note that the function is negative at $x = 1$ but positive at $x = 2$, i.e. $f(1) = -3$ and $f(2) = 3$. So, by Theo. 1.1.1, a real root of Eq. (2.2) lies in the interval (1, 2). Thus, for the points A $(1, -3)$ and B $(2, 3)$ lying on the curve, Eq. (2.1) determines an approximation for the root:

$$x_1 = \{1.f(2) - 2.f(1)\} / \{f(2) - f(1)\} = 3/2 = 1{\cdot}50,$$

where

$$f(x_1) = 27/8 - 3/2 - 3 = -9/8 \approx -1{\cdot}125 < 0.$$

Therefore, again, by Theo. 1.1.1, the real root lies between $x = x_1$ and $x = 2$. Thus, for the points $A_1 \{x_1, f(x_1)\}$ and B $(2, 3)$ lying on the curve, second approximation is similarly found by Eq. (2.1):

$x_2 = \{x_1. f(2) - 2. f(x_1)\} / \{f(2) - f(x_1)\} = (4 \cdot 50 + 2 \cdot 25)/4 \cdot 125 = 1 \cdot 636,$

$$f(x_2) = 1 \cdot 636 \ (0 \cdot 636) \ (2 \cdot 636) - 3 \approx -0 \cdot 254 < 0.$$

Hence, the real root lies between $x = x_2$ and $x = 2$. Thus, for the points $A_2 \{x_2, f(x_2)\}$ and B $(2, 3)$ lying on the curve, third approximation is similarly found by Eq. (2.1):

$$x_3 = \{x_2. f(2) - 2. f(x_2)\} / \{f(2) - f(x_2)\}$$

$$= (4 \cdot 908 + 0 \cdot 508) / 3 \cdot 254 = 1 \cdot 664, \text{ etc. } //$$

§ 3. Newton-Raphson method

Newton's method, also called as the Newton-Raphson method, named after Isaac Newton and Joseph Raphson, is a method for finding successively better approximations to the roots of equation already discovered by earlier methods. Let x_o be an approximate root of Eq. (1.1). If the correct root of the equation is $x_1 = x_o + h$, for some small real number h, so that there holds

$$f(x_1) = f(x_o + h) = 0. \tag{3.1}$$

Expanding the function $f(x_o + h)$, by Taylor's theorem vide Eq. (1.1.4b), above result leads to

$$f(x_o) + h. f'(x_o) + (h^2/2!). f''(x_o) + \ldots + \{(h)^n/n!\}. f^{(n)}(x_o) + \ldots = 0.$$

Accounting the approximations up to the first order of small quantity h, above equation yields

$$h = - f(x_o) / f'(x_o). \tag{3.2}$$

Thus,

$$x_1 = x_o + h = x_o - f(x_o) / f'(x_o) \tag{3.3}$$

is obtained. Successive improved roots are similarly determined by the formula, called *Newton-Raphson formula*

$$x_{n+1} = x_n - f(x_n) / f'(x_n). \tag{3.4}$$

Example 3.1. Using Newton-Raphson method compute a real root of the equation

$$f(x) \equiv x^3 - x - 5 \equiv x(x-1)(x+1) - 5 = 0. \tag{3.5}$$

Solution. For $x = 1$, $f(1) = -5 < 0$ but $f(2) = 1 > 0$. Hence, a real root of the equation lies between 1 and 2. The first order derivative of the function is

$$f'(x) = 3x^2 - 1 \quad \Rightarrow \quad f'(x_n) = 3x_n^2 - 1, \tag{3.6}$$

so that the formula (3.4) determines

$$x_{n+1} = x_n - (x_n^3 - x_n - 5)/(3x_n^2 - 1). \tag{3.7}$$

At the point $x_0 = 1$, above formula (for $n = 0$) gives

$$x_1 = x_0 - (x_0^3 - x_0 - 5)/(3x_0^2 - 1) = 1 + 5/2 = 3 \cdot 5,$$

where

$$f(x_1) = 3 \cdot 5. \, (2 \cdot 5). \, (4 \cdot 5) - 5 = 34 \cdot 375 > 0$$

and

$$f'(x_1) = 3 \, (3 \cdot 5)^2 - 1 = 35 \cdot 75, \quad \text{by Eq. (3.6)}.$$

Therefore, for $n = 1$, Eq. (3.4) determines the second approximation

$$x_2 = x_1 - f(x_1)/f'(x_1) = 3 \cdot 5 - 34 \cdot 375 / 35 \cdot 75 = 2 \cdot 538, \text{ etc. } //$$

§ 4. Problem set

4.1. Using bisection method compute the roots of following equations to the three decimal places:

(i) $x^3 + x^2 + x + 7 = 0$, (ii) $x^3 + x^2 - 1 = 0$, (iii) $x^3 - x^2 - 1 = 0$,

(iv) $x^3 - 3x + 5 = 0$,   (v) $x^3 - 5x + 3 = 0$, (vi) $x^3 - 20 = 0$.

4.2. Using *regula-falsi* method compute the roots of equations in previous problem up to the three decimal places.

4.3. Using Newton-Raphson method compute the roots of following equations to the three decimal places:

(i) $x^3 + 3x^2 - 3 = 0$, (ii) $x^4 + x^2 - 81 = 0$, (iii) $\sin x = x / 2$,

(v) $x^3 - 5x + 3 = 0$, (v) $x + \ln x = 2$,   (vi) $x. \, e^{-2x} = (1/2) \sin x$.

CHAPTER 3

LINEAR PROGRAMMING

§ 1. Introduction

The problems dealing with optimization (i.e. maximization or minimization) of linear functions subject to some linear constraints come within the purview of *linear programming*. Such problems are widely used in industries and in diet problems. The problems of linear programming suiting most to the engineering students are presented in the following. The simplex methods of solving the problems of linear programming are developed. A special class of problems dealing with the transportation and assignment is considered.

Problems of linear programming involving certain variables, relationships amongst them and formulating the objective function and the constraints as considered in the following Example.

Example 1.1. An industry produces two types of models M_1 and M_2. Each M_1 model requires 4 hours for grinding and 2 hours for polishing; whereas each M_2 model requires 2 hours for grinding and 5 hours for polishing. There are 2 grinders and 3 polishers in the industry. Each grinder works for 40 hours a week and each polisher works for 60 hours a week. Each M_1 model earns a profit of \$ 3 and each M_2 model earns \$ 4 profit. How should the production capacity be allocated to the two types of models in order to earn the maximum profit in a week?

Solution. Let x_1 models of the first type (M_1) and x_2 models of the second type (M_2) be produced per week. Then the weekly profit is

$$P = 3x_1 + 4x_2 . \qquad (1.1)$$

The total number of grinding hours needed per week in order to produce above number of models is $4x_1 + 2x_2$, and that cannot exceed the total hours of both the grinders in a week (which is 80). Thus, we have the first constraint:

$$4x_1 + 2x_2 \le 80. \qquad (1.2)$$

Similarly, the total number of polishing hours required per week is $2x_1 + 5x_2$, and that cannot exceed the total capacity of all the three polishers in a week (which is 180 hours):

$$2x_1 + 5x_2 \leq 180. \qquad (1.3)$$

Obviously, there also hold the constraints

$$0 \leq x_1 \qquad \text{and} \qquad 0 \leq x_2. \qquad (1.4)$$

Thus, the problem is to find x_1, x_2 so that P becomes maximum subject to the conditions (1.2) - (1.4). Subtracting relation (1.2) from the second multiple of Eq. (1.3), we get

$$8x_2 \leq 360 - 80 \implies x_2 \leq 35.$$

Accordingly, relation (1.3) yields $x_1 \leq (180 - 5 \times 35)/2 = 2 \cdot 5$. Thus, x_1 being a whole number does not exceed 2. Therefore, the maximum profit in a week can be earned for $x_1 = 2$ and $x_2 = 35$:

$$\text{Max } P = 3 \times 2 + 4 \times 35 = 6 + 140 = 146 \text{ dollars. } //$$

Note 1.1. Elimination of x_2 from (1.2) and (1.3) determines

$$x_1 \leq (400 - 360) / 16 = 2 \cdot 5.$$

Thus, taking maximum integral value 2 for x_1, the relation (1.2) determines the maximum $x_2 = 36$, while relation (1.3) gives max. $x_2 = 35$. Since $x_1 = 2$ and $x_2 = 36$ satisfy relation (1.2) only and not relation (1.3) these values are not acceptable. Instead, $x_1 = 2$ and $x_2 = 35$ satisfy both relations (1.2) and (1.3) maximizing the profit $P = \$ 146$.

Definition 1.1. The variables x_1, x_2 appearing in above problem are called the *decision variables*, the expression (1.1) defines the *objective function* and the relations (1.2) - (1.4) are called the *constraints*.

Example 1.2. (i) A firm making castings melts iron in an electric furnace with the following specifications:

	Minimum	Maximum
Carbon	3·20%	3·40%
Silicon	2·25%	2·35%

(ii) Specifications and costs of various raw materials used for the purpose are as follows:

Material	Carbon	Silicon	Cost (in $)
Steel scrap	0·40	0·15	850 per ton
Cast iron scrap	3·80	2·41	900 "
Re-melt from foundry	3·50	2·35	500 "

(iii) If the total charge of iron metal required is 4 tons, find the weight in kg. of each raw material used in order to make the optimal mix at the minimum cost.

Solution. Let x_1, x_2, x_3 be the amounts (in kg.) of above respective raw materials which minimize the cost P:

$$P = 850\, x_1 / 1000 + 900\, x_2 / 1000 + 500\, x_3 / 1000. \tag{1.5}$$

For iron melt to have a minimum of $3·20\%$ carbon there should hold

$$0·4\, x_1 + 3·8\, x_2 + 3·5\, x_3 \geq 3·2 \times 4000. \tag{1.6}$$

For iron melt to have a maximum of $3·4\%$ carbon there should hold

$$0·4\, x_1 + 3·8\, x_2 + 3·5\, x_3 \leq 3·4 \times 4000. \tag{1.7}$$

For iron melt to have a minimum of $2·25\%$ silicon there should hold

$$0·15\, x_1 + 2·41\, x_2 + 2·35\, x_3 \geq 2·25 \times 4000. \tag{1.8}$$

For iron melt to have a maximum of $2·35\%$ silicon there should hold

$$0·15\, x_1 + 2·41\, x_2 + 2·35\, x_3 \leq 2·35 \times 4000. \tag{1.9}$$

Also, the total material is
$$x_1 + x_2 + x_3 = 4000; \tag{1.10}$$

and there also hold the constraints

$$0 \leq x_1, \quad 0 \leq x_2, \quad 0 \leq x_3. \tag{1.11}$$

Thus, we have to evaluate x_1, x_2, x_3 making P minimum and satisfying the conditions (1.6) - (1.11).

Putting for x_3 from Eq. (1.10) and rewriting the conditions (1.6) - (1.9) as

$$4\,x_1 + 38\,x_2 + 35\,(4000 - x_1 - x_2) \geq 128{,}000$$

$\Rightarrow$

$$31x_1 - 3x_2 \leq 12{,}000 \quad (1.6a); \quad 31x_1 - 3x_2 \geq 4{,}000, \quad (1.7a)$$

$$15\,x_1 + 241\,x_2 + 235\,(4000 - x_1 - x_2) \geq 900{,}000$$

$\Rightarrow$

$$110\,x_1 - 3x_2 \leq 20{,}000 \quad (1.8a); \quad \text{and} \quad 110\,x_1 - 3x_2 \geq 0. \quad (1.9a)$$

Solving Eqs. (1.6a) and (1.8a), we obtain $x_1 \leq 8000 / 79 = 101{\cdot}266$ and

$$x_2 \leq (110/3)\,x_1 \leq (110/3)\,(8000/79) = 880{,}000/237.$$

Consequently, Eq. (1.10) determines

$$x_3 \geq 4{,}000 - (8{,}000/79 + 880{,}000/237) = 44{,}000/237.$$

These values make P minimum:

$$P = (850 \times 8 + 900 \times 880/3 + 500 \times 44/3)/79 = 3520{\cdot}67. \; //$$

§ 2. Graphical method

Linear programming problem involving two variables can also be solved by a graphical technique. We sketch the graph of constraints given in Example 1.1 and solve the problem in the following.

Example 2.1. Solve the problem enunciated in Example 1.1 by graphical method.

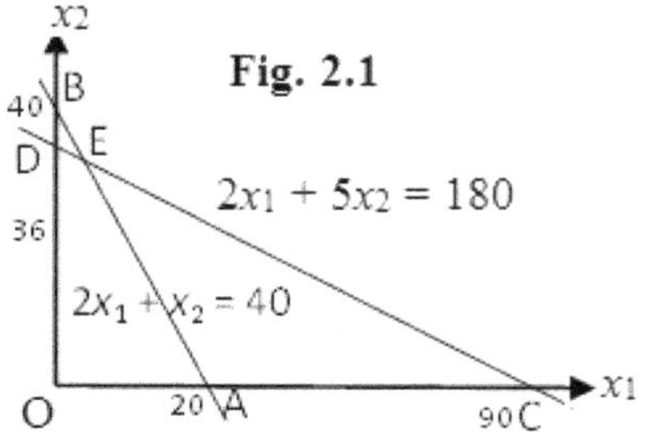

Solution. We draw the graphs of straight lines

Fig. 2.1

$$4x_1 + 2x_2 = 80, \quad 2x_1 + 5x_2 = 180,$$

i.e.

$$x_1/20 + x_2/40 = 1, \quad (2.1)$$

and

$$x_1/90 + x_2/36 = 1. \quad (2.2)$$

The points on or below both the lines satisfy the conditions (1.2) and

(1.3). Also, Eq. (1.4) ascertains these points in the first quadrant only. Thus, they lie in the shaded region OAED. The values of x_1, x_2 and corresponding P at the vertices of this region are

Point	x_1	x_2	P
O	0	0	0
A	20	0	60
E	2·5	35	147·5
D	0	36	144

Thus, $x_1 = 2·5$ and $x_2 = 35$ maximize $P = 147·5$. //

Example 2.2. Maximize the objective function

$$P(x, y) = 2x + 3y, \qquad (2.3)$$

subject to the constraints

$$x + y \leq 30, \qquad (2.4)$$

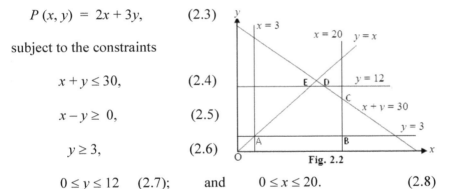

$$x - y \geq 0, \qquad (2.5)$$

$$y \geq 3, \qquad (2.6)$$

Fig. 2.2

$$0 \leq y \leq 12 \quad (2.7); \qquad \text{and} \qquad 0 \leq x \leq 20. \qquad (2.8)$$

Solution. The relations (2.7) and (2.8) imply that either of x and y lie in the first quadrant only. Further, for (2.5) and (2.6) the last two relations also imply

$$3 \leq y \leq 12 \qquad (2.9); \text{ and} \qquad 3 \leq x \leq 20. \qquad (2.10)$$

The relevant constraint relations are as per relations (2.4), (2.9) and (2.10). Sketching the graphs of the straight lines

$$x + y = 30, \qquad 3 = y, \qquad y = 12, \quad 3 = x \quad \text{and} \quad x = 20,$$

their points of intersection are A (3,3), B (20,3), C (20,10), D (18,12) and E (12,12). Thus, the constraints are satisfied for the convex region ABCDE (shaded in the Fig. 2.2). In the following, we compute the objective function at the vertices of this region.

Point	x	y	$P(x, y)$
A	3	3	$2 \times 3 + 3 \times 3 = 15$
B	20	3	$2 \times 20 + 3 \times 3 = 49$
C	20	10	$2 \times 20 + 3 \times 10 = 70$
D	18	12	$2 \times 18 + 3 \times 12 = 72$
E	12	12	$2 \times 12 + 3 \times 12 = 60$

Therefore, the maximum value of P is 72 at the point D (18, 12). //

Example 2.3. A company manufactures two types of cloth using three different colours of wool. One metre length of type A cloth requires 120 gms. of red wool, 150 gms. of green wool and 90 gms. of yellow wool. On the other hand, one metre of type B cloth requires 150 gms. of red wool, 60 gms. of green wool and 240 gms. of yellow wool. The manufacturer has 30 kg. of red wool, 30 kg. of green wool and 36 kg. of yellow wool. The profit made by the manufacturer on one metre of cloth A is $ 5 and on one metre of cloth B is $ 3. Find the best combination of the quantities of both types of cloth in order to earn the maximum profit.

Solution. Let x metres of cloth A and y metres of cloth B give maximum profit

$$P(x, y) = 5x + 3y. \qquad (2.11)$$

The quantity of red wool required is

$$120x + 150y \leq 30,000, \text{ i.e.} \quad 4x + 5y \leq 1,000. \qquad (2.12)$$

Green wool required is

$$150x + 60y \leq 30,000, \quad \text{i.e.} \quad 5x + 2y \leq 1,000; \qquad (2.13)$$

and yellow wool required is

$$90\,x + 240\,y \leq 36{,}000,$$

i.e.

$$3x + 8y \leq 1{,}200. \qquad (2.14)$$

Also, there hold $0 \leq x$ and $0 \leq y$. Solving Eqs. (2.12) and (2.13) we get $x \leq 3000 / 17$ and $y \leq 1000/17$. These values also satisfy relations (2.14). Thus, the maximum profit is

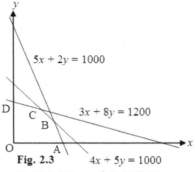

Fig. 2.3

$$P = (5 \times 3000 + 3 \times 1000) / 17 = 18{,}000 /17 = \$\,1{,}058{\cdot}82.$$

Alternately: Drawing the graphs of straight lines

$$4x + 5y = 1{,}000, \qquad \text{i.e.} \qquad x / 250 + y / 200 = 1;$$

$$5x + 2y = 1{,}000, \qquad \text{i.e.} \qquad x / 200 + y / 500 = 1;$$

$$3x + 8y = 1{,}200, \qquad \text{i.e.} \qquad x / 400 + y / 150 = 1;$$

we get graph of relevant function by the shaded portion OABCD. We computing coordinates of the vertices of this region: O (0, 0), A (200, 0), B (3000/17, 1000/17), C (2000/17, 1800/17), and D (0, 150). Accordingly, the function at these points has values:

$$P_O = 0, \quad P_A = 1{,}000, \quad P_B = 1{,}8000/17, \quad P_C = 15{,}400/17, \quad P_D = 450;$$

out of which $P_B = 1{,}8000 / 17 = 1{,}058{\cdot}82$ is maximum. //

Example 2.4. A cold drinks factory has two plants located at towns T_1 and T_2. Each plant produces three different types of drinks A, B, C. The production capacity of the plants per day is as follows:

Cold drinks	Plant at T_1	Plant at T_2
A	6,000 bottles	2,000 bottles
B	1,000 bottles	2,500 bottles
C	3,000 bottles	3,000 bottles

The marketing department of the factory forecasts a demand of 80,000 bottles of A, 22,000 bottles of B and 40,000 bottles of drink C in the month of August. The operating costs per day of plants at T_1, T_2 are $ 6,000 and $ 4,000 respectively. Find the number of days for which each plant must be run in August so as to minimize the operating costs in meeting the market demand.

Solution. Let the plants at T_1, T_2 be run for x and y days respectively in August. So, the operating costs are

$P (x, y) = 6000 x + 4000 y.$ (2.15)

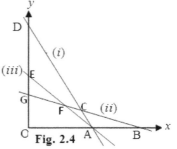

Constraints on the demand for three types of drinks are:

(i) for A: $6000 x + 2000 y \geq 80,000$

$\Rightarrow$

$3x + y \geq 40,$ (2.16)

Fig. 2.4

(ii) for B: $1000 x + 2500 y \geq 22,000 \Rightarrow 2x + 5y \geq 44,$ (2.17)

(iii) for C: $3000 x + 3000 y \geq 40,000 \Rightarrow 3 (x + y) \geq 40,$ (2.18)

Also, there hold $0 \leq x$ and $0 \leq y$. Solving Eqs. (2.16) and (2.17) we get $x \geq 12$ and $y \geq 4$; which also satisfy the constraint (2.18). These values of x and y determine the minimum operating cost

$$P = 6000 \times 12 + 4000 \times 4 = \$ 88,000.$$

Alternately, drawing the graphs of equations

$3x + y = 40,$ $2x + 5y = 44$ and $3 (x + y) = 40,$ (2.19)

the common solutions of all the constraints (2.16) - (2.18) lie on the right side of all three straight lines. Thus, three vertices of the shaded portion are B (22, 0), C (12, 4) and D (0, 40) where the operating costs are

$P_B = 6000 \times 22 = 132,000,$ $P_C = 88,000$ and $P_D = 4000 \times 40 = 160,000.$

Thus, the vertex C provides the minimum operating cost $ 88,000 for running the plant at T_1 for 12 days and that at T_2 for 4 days. //

§ 3. Some exceptional cases

The constraints generally give region of feasible solution that may be bounded or unbounded. In the problems involving two variables and a finite solution as observed in the preceding section the optimal solution exists at a vertex of the feasible region. Thus, if there exists an optimal solution for a linear programming problem it will be at one of the vertices of the solution space. In the previous section we had problems with a unique solution. But, it is not always so. In fact, there may exist any of the following possibilities for the solution of a linear programming problem (LPP):

(*i*) a unique optimal solution,

(*ii*) an infinite number of optimal solutions,

(*iii*) an unbounded solution, (*iv*) no solution at all.

In the following we discuss some examples with these exceptional cases.

Example 3.1. A firm uses milling machines, grinding machines and lathes to produce two motor parts. The machining times required for each part, the machining times available on different machines and the profit on each motor part are given below. Determine the number of parts I and II to be manufactured per week to maximize the profit.

Type of machine	Machining time required (in minutes) for		Maximum time available per week (in minutes)
	I part	II part	
Milling machines	10	4	2,000
Grinding machines	3	2	900
Lathes	6	12	3,000
Profit per unit (in $)	100	40	

Solution. Let x and y be the number of parts I and II respectively manufactured per week. The objective function (maximizing the profit) is

$$P(x, y) \equiv 100\, x + 40\, y. \qquad (3.1)$$

Constraints for time available on each machine are:

  (*i*) for milling machines: $10\,x + 4\,y \le 2000,$ (3.2)

  (*ii*) for grinding machines: $3\,x + 2\,y \le 900,$ (3.3)

  (*iii*) for lathes: $6\,x + 12\,y \le 3000,$ (3.4)

Also,

$$x, y \ge 0. \qquad (3.5)$$

Solving Eqs. (3.2) and (3.3) we get $x \le 50$ and $y \le 375$ which do not satisfy relation (3.4). Similarly, the solution of Eqs. (3.3) and (3.4): $x \le 200$ and $y \le 150$ also do not satisfy relation (3.2). However, the common solution of Eqs. (3.2) and (3.4):

$$x \le 125, \qquad y \le 187{\cdot}5 \qquad (3.6)$$

does satisfy relation (3.3) also. So, relations (3.6) gives a solution meeting all the constraints (3.2) - (3.5) for which the maximum profit function is

$$P\,(125,\ 187{\cdot}5) = 12{,}500 + 7{,}500 \ = \ \$\ 20{,}000. \ //$$

Note 3.1. It is interesting to note here another solution: $x = 200$, $y = 0$ satisfying all the constraints (3.2) - (3.5) and giving the same (maximum) value of $P = 20{,}000$.

Alternately, solving the problem by graphical method we first draw the graphs of equations

  $5\,x + 2\,y = 1000$ (3.2a); $3\,x + 2\,y = 900,$ (3.3a)
and
  $x + 2\,y = 500.$ (3.4a)

For constraints (3.5), both x and y lie in the first quadrant and the shaded portion and the shaded portion in the Fig. 3.1 forms the solution space. The vertices of the solution space have coordinates O (0, 0), A(200, 0), B (125, 187·5) and C (0, 250) where the profit function is

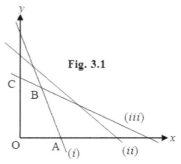

Fig. 3.1

$$P_O = 0; \quad P_A = 20{,}000; \quad P_B = 20{,}000; \quad P_C = 10{,}000.$$

Thus, the profit is maximum at both points A and B. Indeed, any values of x and y lying on the line AB meet all the constraints $(3.2) - (3.5)$ and give the same (maximum) profit. As such, there are infinite points on the line AB providing the same (maximum) profit. //

§ 4. General linear programming problem

Any LPP involving more than two variables may be expressed as follows:

To find the values of variables x_i, $i = 1, 2, \ldots, n$, which maximize (or minimize) the objective function

$$P = c_1 x_1 + c_2 x_2 + \ldots + c_n x_n = \sum_{i=1}^{n} c_i x_i, \qquad (4.1)$$

subject to the constraints

$$\left.\begin{array}{l} a_{11} x_1 + a_{12} x_2 + \ldots + a_{1n} x_n \leq b_1, \\[4pt] a_{21} x_1 + a_{22} x_2 + \ldots + a_{2n} x_n \leq b_2, \\[4pt] \ldots\ldots\ldots\ldots\ldots\ldots\ldots\ldots\ldots\ldots\ldots, \\[4pt] a_{m1} x_1 + a_{m2} x_2 + \ldots + a_{mn} x_n \leq b_m, \end{array}\right\} \qquad (4.2)$$

and the non-negative restrictions

$$x_1, x_2, \ldots, x_n \geq 0. \qquad (4.3)$$

Definition 4.1. (i) A set of values $x_1, x_2, \ldots, x_n$ satisfying the constraints of the LPP forms a *solution*.

(ii) Any solution to a LPP satisfying the non-negative restrictions of the problem is called a *feasible solution*.

(iii) A feasible solution maximizing (or minimizing) the objective function of the LPP is called the *optimal solution* of the problem.

Definition 4.2. If the constraints of a LPP satisfy constraints (4.2) then the non-negative variables s_i, $i = 1, 2, \ldots, m$, given by

$$\sum_{j=1}^{n} a_{ij} x_j + s_i = b_i, \qquad i = 1, 2, \ldots, m; \qquad (4.4)$$

are called the *slack variables*.

Definition 4.3. If the constraints of a LPP satisfy

$$\sum_{j=1}^{n} a_{ij} x_j \geq b_i, \qquad i = 1, 2, \ldots, m; \tag{4.5}$$

then the non-negative variables $\bar{s}_i$, $i = 1, 2, \ldots, m$, given by

$$\sum_{j=1}^{n} a_{ij} x_j - \bar{s}_i = b_i, \qquad i = 1, 2, \ldots, m; \tag{4.6}$$

are called the *surplus variables*.

4.1. Canonical and standard forms

Definition 4.4. When the constraints of a LPP satisfy constraints (4.2) the form of LPP is called *canonical form* and it has the following characteristics:

(*i*) The objective function is to be maximized;

(*ii*) All constraints are of type $\leq$ (less or equal to);

(*iii*) All variables are non-negative.

On the other hand, if the constraints (4.2) have equations (and not inequalities) the form of LPP is called *standard form*. Such a form has the following characteristics:

(*i*) The objective function is to be maximized;

(*ii*) Each b_i, $i = 1, 2, \ldots, m$, is non-negative;

(*iii*) All variables x_i's are non-negative.

(*iv*) All constraints are expressed as equations.

Note 4.1. Minimizing a function P given by Eq. (4.1) is equivalent to maximize its negative, viz.

$$Q \equiv -P = -c_1 x_1 - c_2 x_2 - \ldots - c_n x_n. \tag{4.7}$$

Example 4.1. Convert the following LPP to the standard form:

Maximize $P = 3x + 5y + 7z$ subject to the constraints

$$6x - 4y \leq 5; \quad 3x + 2y + 5z \geq 11; \qquad 4x + 3z \leq 2; \qquad x, y \geq 0.$$

Solution. As the third variable z is unrestricted we put

$$z = z_1 - z_2, \tag{4.8}$$

where both $z_1, z_2 \geq 0$. Accordingly, the given constraints can be restated as

$$6x - 4y \leq 5; \qquad 3x + 2y + 5z_1 - 5z_2 \geq 11;$$

$$4x + 3z_1 - 3z_2 \leq 2; \quad x, y, z_1, z_2 \geq 0.$$

Introducing slack / surplus variables, the problem in standard form becomes:

Maximize $P = 3x + 5y + 7z_1 - 7z_2$ subject to the constraints

$$6x - 4y + s_1 = 5; \qquad 3x + 2y + 5z_1 - 5z_2 - s_2 = 11;$$

$$4x + 3z_1 - 3z_2 + s_3 = 2; \qquad x, y, z_1, z_2, s_1, s_2, s_3 \geq 0. \; //$$

Example 4.2. Express the following problem in the standard form:

Minimize $P = 3x + 4y$ subject to the constraints

$$2x - y - 3z = -4; \quad 3x + 5y + u = 10; \qquad x - 4y = 12; \, x, z, u \geq 0.$$

Solution. Since y is unrestricted, we put $y = y_1 - y_2$; $y_1, y_2 \geq 0$. The problem is restated in the standard form as:

Maximize $Q = -P = -3x - 4y_1 + 4y_2$ subject to the constraints

$$-2x + y_1 - y_2 + 3z = 4; \qquad 3x + 5y_1 - 5y_2 + u = 10;$$

$$x - 4y_1 + 4y_2 = 12; \quad x, y_1, y_2, z, u \geq 0. \; //$$

§ 5. Simplex method

5.1. While solving an LPP by graphical method the region of feasible solutions is seen as convex, bounded by vertices and edges joining them. Some vertex provided the optimal solution. If the optimal solution was not unique the optimal points were on an edge. The same was also true for a general LPP. Truly speaking the main problem is locating the particular vertex of the convex region providing the optimal solution. The most commonly method used for locating the optimal vertex is the simplex method, i.e. moving step by step from one vertex to the adjacent one. Amongst all adjacent vertices, the one providing the better value of the objective function over that at the preceding vertex is then repeated. The number of vertices being finite a finite number of steps leads to an optimal vertex.

We have the following definitions:

Definition 5.1. (*Solution*) The variables x_i, $i = 1, 2, \ldots , n$ give a solution of a general LPP if they satisfy the constraints vide Eqs. (4.2).

Definition 5.2. (*Feasible solution*) The solution given by x_i's is called a feasible solution if the non-negativity restrictions vide Eqs. (4.3) are also met. A set S of all feasible solutions is called the *feasible region*. If S is empty a linear programme is said to be *infeasible*.

Definition 5.3. (*Basic solution*) A solution given by m basic variables, where each of the non-basic variable(s) are equated to zero, is called the basic solution.

Definition 5.4. (*Basic feasible solution*) The basic solution satisfying the non-negativity restrictions vide Eqs. (4.3) is called the basic feasible solution.

Definition 5.5. (*Optimal solution*) The basic feasible solution optimizing the objective function is called the optimal solution.

Definition 5.6. (*Non-degenerate basic feasible solution*) The basic feasible solution containing exactly m non-zero (indeed positive) basic variables is called non-degenerate basic feasible solution.

If any of the basic variables becomes zero the solution is called the *degenerate basic feasible solution*.

5.2. An infinite number of solutions is reduced to a finite number of promising solutions by using the following points:

(i) When an LPP involves m constraints and $m + n$ variables ($m \leq n$) the starting solution is obtained by making n variables zero and then solving the remaining m equations for m unknowns. Thus, there exists a unique solution provided the equations are solvable. The n zero variables are called the *non-basic variables* while the remaining m variables as *basic variables*. The basic variables form a *basic solution*.

(ii) Usually the variables in an LPP must always be non-negative. However, some of the basic solutions may contain negative variables forming *basic infeasible solutions*.

These solutions should be discarded. To obtain this, we begin with a non-negative basic solution. The next basic solution must also be non-negative which is ensured by the feasibility condition. Such a solution is called a *basic feasible solution*. Thus, a basic feasible solution satisfies both: constraints and non-negativity of the variables.

In case the basic feasible solution has all positive variables then the solution is *non-degenerate*. But, if some of the variables in the solution are zero it is called *degenerate solution*.

(iii) Equating one of the basic variables in a basic feasible solution to zero and replacing it by a new non-basic variable, a new basic feasible solution can be obtained. The eliminated variable is termed as the *outgoing variable* and the new variable is called the *incoming variable*. The incoming variable is supposed to improve the objective function, which is ensured by the feasible condition. The process is repeated till no further improvement is possible. The resulting solution is called the *optimal basic feasible solution* or briefly *optimal solution*.

5.3. The simplex method is, thus, based on the following conditions:

(i) Feasibility condition: It ensures the subsequent solution to be basic feasible provided the starting solution is basic feasible.

(ii) Optimality condition: It ensures that the improved solutions will be obtained.

Example 5.1. Find all the basic solutions of the following system of linear equations identifying in each case the basic and non-basic variables:

$$2x + y + 4z = 11, \quad 3x + y + 5z = 14. \tag{5.1}$$

Investigate whether the basic solutions are degenerate basic solutions or not. Hence, find the basic feasible solution of the system.

Solution. Since there are two constraints involving three variables in the problem, a basic solution can be obtained by setting any one variable zero; and then solving the equations for remaining (basic) variables. Thus, the total number of basic solutions is $^3C_2 = 3$. The basic solutions possess the following characteristics:

Basic solns.	Basic variables	Non-basic variables	Values of basic variables	Feasible soln.	Degenerate soln.
1.	x, y	$z (= 0)$	$2x + y = 11,$ $3x + y = 14$ $\Rightarrow x = 3, y = 5$	yes	no
2.	y, z	$x (= 0)$	$y + 4z = 11,$ $y + 5z = 14$ $\Rightarrow y = -1, z = 3$	no	no
3.	x, z	$y (= 0)$	$2x + 4z = 11,$ $3x + 5z = 14$ $\Rightarrow x = \cdot 5, z = 2 \cdot 5$	yes	no

Thus, the basic feasible solutions are at Sr. Nos. 1 and 3. These solutions are non-degenerate as well. //

Example 5.2. Find an optimal solution of the following LPP by computing all basic solutions and then choosing the one maximizing the objective function $P = 2x + 3y + 4z + 7u$ subject to the constraints

$$2x + 3y - z + 4u = 8, \quad x - 2y + 6z - 7u = -3, \ x, y, z, u \geq 0.$$

Solution. Since there are two (main) constraints and four variables, a basic solution can be obtained by setting any two variables zero. The total number of such basic solutions will be $^4C_2 = 6$. The basic solutions possess the following characteristics:

Basic soln.	Basic variables	Non-basic Variables	Values of basic variables	Feasible soln.	Value of P	Optimal soln.
1.	x, y	$z,$ $u\,(=0)$	$2x + 3y = 8,$ $x - 2y = -3$ $\Rightarrow x = 1, y = 2$	yes	8	no
2.	x, z	$y,$ $u\,(=0)$	$2x - z = 8,$ $x + 6z = -3$ $\Rightarrow x = 45/13,$ $z = -14/13$	no	-	-
3.	x, u	$y,$ $z\,(=0)$	$x + 2u = 4,$ $x - 7u = -3$ $\Rightarrow x = 22/9,$ $u = 7/9$	yes	10·3	no
4.	y, z	$x,$ $u\,(=0)$	$3y - z = 8,$ $-2y + 6z = -3$ $\Rightarrow y = 45/16,$ $z = 7/16$	yes	10·2	no
5.	y, u	$x,$ $z\,(=0)$	$3y + 4u = 8,$ $2y + 7u = 3$ $\Rightarrow y = 44/13,$ $u = -7/13$	no	-	-
6.	z, u	$x,$ $y\,(=0)$	$-z + 4u = 8,$ $6z - 7u = -3$ $\Rightarrow z = 44/17,$ $u = 45/17$	yes	28·9	yes

Thus, the optimal basic feasible solution is at the last serial number:

$$x = 0, \qquad y = 0, \qquad z = 44/17, \qquad u = 45/17;$$

giving the maximum value of $P = 28 \cdot 9$. //

§ 6. Working procedure for the simplex method

If there exists an initial basic feasible solution, an optimal solution to any LPP by simplex method is found as per the following procedure:

Step 1. Check if the objective function P is to be maximized or minimized. In the latter case, convert it to the problem of maximizing its negative, i.e. $-P$.

Step 2. Check if all b's (appearing in the constraints (4.2)) are positive. If any of these is negative multiply both sides of the concerned constraint by -1 so as to make its right side positive.

Step 3. Express the LPP in the standard form by converting all inequalities of constraints into equations by introducing (if necessary) some slack / surplus variables.

Step 4. Find an initial basic feasible solution.

Step 5. Apply optimality test.

Step 6. Identify the incoming and outgoing variables.

Example 6.1. Using simplex method maximize $P = 5x + 3y$ subject to the constraints

$$x + y \leq 2, \quad 5x + 2y \leq 10, \quad 3x + 8y \leq 12, \quad x, y \geq 0.$$

Solution. Steps 1 and 2 are not necessary as their requirements are fully met.

Step 3. Expressing the problem in the standard form:

Maximize $P = 5x + 3y + 0s_1 + 0s_2 + 0s_3$ subject to the constraints

$$x + y + s_1 + 0s_2 + 0s_3 = 2 \quad (6.1); \quad 5x + 2y + 0s_1 + s_2 + 0s_3 = 10, \quad (6.2)$$

$$3x + 8y + 0s_1 + 0s_2 + s_3 = 12 \quad (6.3); \qquad x, y, s_1, s_2, s_3 \geq 0. \quad (6.4)$$

Step 4. To find an initial basic feasible solution we proceed as follows. There are three equations involving five variables. So, their basic solutions can be obtained by assigning zero values to any two of the variables. Setting $x = y = 0$ (holding at origin in the graphical method for solution) a basic solution is

$$x = y = 0, \quad s_1 = 2, \quad s_2 = 10, \quad s_3 = 12. \tag{6.5}$$

All s_i's, $i = 1, 2, 3$, being positive the basic solution (formed by the basic variables s_i's and non-basic variables x, y) is feasible and non-degenerate. However, the value of P for this solution is zero (which is not optimal). In the following, we workout all basic solutions and compute the values of P:

x	y	s_1	s_2	s_3	P	Feasible soln.
0	0	2	10	12	0	yes
0	2	0	6	-4	6	no
0	5	-3	0	-28	15	no
0	1·5	·5	7	0		yes
2	0	0	0	6	10	yes
2	0	0	0	6	10	yes
4	0	-2	-10	0	20	no
2	0	0	0	6	10	yes
·8	1·2	0	3·6	0	7·6	yes
28/17	15/17	$-9/17$	0	0	10·9	no

Thus, the optimal feasible solution is $x = 2$, $y = 0$ giving the maximum value 10 of P. //

Example 6.2. A firm manufactures three different products, which are processed on three different types of machines. The processing data and the profit per unit of products A, B, C is according to the following table:

Machine	Time taken by machine Per unit (in minutes) Products			Daily capacity of Machine (in minutes)
	A	B	C	
M_1	2	3	2	440
M_2	4	-	3	470
M_3	2	5	-	430
Profit (in $)	4	3	6	

Determine the daily number of units to be manufactured for each product in order to maximize the profit. Presume that all units manufactured are sold in the market.

Solution. Let x, y, z be the number of units of each product A, B, C respectively manufactured daily in order to maximize the profit. Thus, formulating the LPP as:

Maximize $P = 4x + 3y + 6z$ subject to the constraints:

(i) for machine M_1: $2x + 3y + 2z \leq 440,$ (6.6)

(ii) for machine M_2: $4x + 0y + 3z \leq 470,$ (6.7)

(iii) for machine M_3: $2x + 5y + 0z \leq 430,$ (6.8)
and

$$x, y, z \geq 0.$$ (6.9)

Thus, we note that the Steps 1 and 2 of the working procedure are fully satisfied.

Step 3. Introducing three slack variables s_1, s_2, s_3 the problem is converted into the standard form:

To maximize $P = 4x + 3y + 6z + 0s_1 + 0s_2 + 0s_3$ subject to constraints

$$2x + 3y + 2z + s_1 + 0s_2 + 0s_3 = 440,$$ (6.6a)

$$4x + 0y + 3z + 0s_1 + s_2 + 0s_3 = 470, \qquad (6.7a)$$

$$2x + 5y + 0z + 0s_1 + 0s_2 + s_3 = 430, \qquad (6.8a)$$

and

$$x, y, z, s_1, s_2, s_3 \geq 0. \qquad (6.9a)$$

Step 4. There being three equations containing six variables basic solutions can be obtained by assigning zero values to any three variables. A basic non-degenerate feasible solution is

$x = y = z = 0$ (non-basic soln.), $s_1 = 440$, $s_2 = 470$, $s_3 = 430$ (basic soln.)

However, the value of P for this solution is zero (and not maximum). Similarly, other basic solutions are found as per the following table.

x	y	z	s_1	s_2	s_3	P	Feasible soln.
0	0	0	440	470	430	0	yes
0	0	220	0	− 190	430	--	no
0	0	470/3	380/3	0	430	940	yes
0	440/3	0	0	470	− 910/3	--	no
0	86	0	182	470	0	258	yes
0	380/9	470/3	0	0	1970/9	1066·67	yes
0	86	91	0	197	0	804	yes
0	86	470/3	− 394/3	0	0	--	no
220	0	0	0	− 410	− 10	--	no
235/2	0	0	205	0	195	470	yes
215	0	0	10	− 390	0	--	no
−190	0	410	0	0	810	--	no

215	0	5	0	− 405	0	--	no
215	0	−130	270	0	0	--	no
235/2	205/3	0	0	0	− 440/3	--	no
455/2	− 5	0	0	− 440	0	--	no
235/2	39	0	88	0	0	587	yes
985/14	405/7	440/7	0	0	0	832·14	yes

Thus, the optimal feasible solution is $x = 0$, $y = 380/9$, $z = 470/3$ giving the maximum profit $P = \$ 1066·67$. Since x, y, z are whole numbers the acceptable optimal solution is $x = 0$, $y = 42$, $z = 156$ earning the maximum profit $P = \$ 1062$. //

§ 7. Problem set

7.1. Reduce the following problem to the standard form:

Determine $x \geq 0, y \geq 0, z \geq 0$ so as to maximize $P = 3x + 5y + 8z$ subject to the constraints

$$2x - 5y \leq 6, \quad 3x + 2y + z \geq 5, \quad 3x + 4z \leq 3.$$

7.2. Express the following LPP to the standard form:

Minimize $P = 3x + 2y + 5z$ subject to the constraints

$$- 5x + 2y \leq 5, \quad 2x + 3y + 4z \geq 7, \quad 2x + 5z \leq 3, \quad x, y, z \geq 0.$$

7.3. Convert the following LPP to the standard form:

Maximize $P = 3x - 2y + 4z$ subject to the constraints

$$x + 2y + z \leq 8, \quad 2x - y + z \geq 2, \quad 4x - 2y - 3z = - 6, \quad x, y \geq 0.$$

7.4. Obtain all the basic solutions to the following system of linear equations:
$$x + 2y + z = 4, \qquad\qquad 2x + y + 5z = 5.$$

7.5. Show that the following system of linear equations has two degenerate feasible basic solutions and the non-degenerate basic solution is not feasible:

$$2x + y - z = 2, \qquad 3x + 2y + z = 3.$$

7.6. Find all the basic solutions to the following problem:

Maximize $P = x + 3y + 3z$ subject to the constraints

$$x + 2y + 3z = 4, \quad 2x + 3y + 5z = 7, \text{ and } \quad x \geq 0, y \geq 0, z \geq 0.$$

Which of the basic solutions are: (*i*) non-degenerate basic feasible, (*ii*) optimal basic feasible?

CHAPTER 4

NUMERICAL SOLUTIONS OF ORDINARY DIFFERENTIAL EQUATIONS

§ 1. Introduction

We consider a first order differential equation

$$dy / dx = f(x, y) \tag{1.1}$$

with the initial condition

$$y_0 = y(x_0), \tag{1.2}$$

i.e. y_0 is a solution of the differential equation at $x = x_0$. There are different methods to solve the equation. Those of Picard and Taylor yield solutions for y as a power series in x from which the values of y can be evaluated by direct substitution. On the other hand, the methods of Euler, Runge-Kutta, Milne, Adams-Bashforth etc. give a set of values of x and y. In the latter methods the values of y are calculated for equal intervals of x. As such, these are also called *step-by-step methods*.

Euler and Runge-Kutta methods are employed for computing y over a limited range of x-values. On the other hand, Milne and Adams-Bashforth methods are used for y over a wide range of x-values. These latter methods require initial values obtainable by Picard's method or by Taylor series or Runge-Kutta methods.

The initial condition (1.2) is specified at the point x_0. The problems wherein all the initial conditions are given at the initial point only are called *initial value problems*. However, there are also problems (involving second and higher order differential equations) where the conditions may be given at two (or more) points. Such problems are called *boundary value problems*. We deal with both types of these (initial and boundary value) problems for second order differential equations in the present chapter.

§ 2. Picard's method for the solution of Eq. (1.1)

Integrating Eq. (1.1) between the limits $P_0(x_0, y_0)$ to $P(x, y)$:

$$\int_{y_0}^{y} dy = \int_{x_0}^{x} f(x, y) \, dx \tag{2.1}$$

$\Rightarrow$

$$y = y_0 + \int_{x_0}^{x} f(x, y)\, dx. \qquad (2.2)$$

Choosing $y = y_0$ in $f(x, y)$ the first approximation y_1 to the solution y is thus given by

$$y_1 = y_0 + \int_{x_0}^{x} f(x, y_0)\, dx. \qquad (2.3)$$

Next, choosing $y = y_1$ in $f(x, y)$ the second approximation y_2 for the solution y is given by

$$y_2 = y_0 + \int_{x_0}^{x} f(x, y_1)\, dx. \qquad (2.4)$$

Similarly, a third approximation y_3 is

$$y_3 = y_0 + \int_{x_0}^{x} f(x, y_2)\, dx. \qquad (2.5)$$

Continuing the process a sequence of functions $y_1, y_2, y_3, \ldots$ of x is obtained each giving a better approximation of the solution than the preceding one.

Note 2.1. Picard's method is applicable to the equations where the successive integrations can be performed easily.

§ 3. Taylor's series method for the solution of Eq. (1.1)

Differentiation of Eq. (1.1) with respect to x yields

$$d^2 y / dx^2 = df / dx = \partial f / \partial x + (\partial f / \partial y).(dy / dx),$$

i.e.

$$y'' = f_x + f_y \cdot f, \qquad (3.1)$$

where the primes denote differentiation w.r.t. x and suffixes are used to denote partial differentiation.

Differentiating Eq. (3.1) successively w.r.t. x the derivatives y''', $y^{(iv)}$, etc. can also be evaluated. Computing the values of these derivatives at the point P_0 (x_0, y_0) we find $y'(0)$, $y''(0)$, $y'''(0)$, etc. Hence, the Taylor's series for y about x_0 is

$$y = y_0 + (x - x_0)\, y'(0) + \{(x - x_0)^2 / 2!\}\, y''(0)$$

$$+ \{(x - x_0)^3 / 3!\} \, y''' \, (0) + \ldots \, \infty, \tag{3.2}$$

determining y for every value of x for which Eq. (3.2) converges. Let y_1 obtainable from Eq. (3.2) be the value of y at $x = x_1$:

$$y_1 \; = \; y_0 + (x_1 - x_0) \, y' \, (0) + \{(x_1 - x_0)^2 / 2!\} \, y'' \, (0) + \ldots \, \infty, \tag{3.3}$$

Further, $y' \, (= dy/dx = f)$, y'' can be evaluated at $x = x_1$ from Eqs. (1.1) and (3.1) respectively. Thereafter, y can be expanded by Taylor's theorem about $x = x_1$:

$$y \; = \; y_1 + (x - x_1) \, y' \, (1) + \{(x - x_1)^2 / 2!\} \, y'' \, (1)$$

$$+ \{(x - x_1)^3 / 3!\} \, y''' \, (1) + \ldots \, \infty, \tag{3.4}$$

Example 3.1. Apply Taylor's series method to evaluate y at $x = 0 \cdot 1$ and $x = 0 \cdot 2$ to five places of decimals from

$$dy \, / \, dx \; = \; x^2 \, y - 1 \tag{3.5}$$

together with

$$y_0 = y \, (0) \; = \; 1. \tag{3.6}$$

Solution. For Eqs. (3.5) and (3.6) we have

$$y' \; = \; dy \, / \, dx \; = \; x^2 \, y - 1 \quad \Rightarrow \quad y' \, (0) \; = \; (0)^2 . y \, (0) - 1 = - \, 1.$$

Differentiating Eq. (3.5) successively and evaluating the derivatives at $x = 0$:

$$y'' \; = 2xy + x^2 y' \quad \Rightarrow \quad y''(0) = 0;$$

$$y''' = 2y + 4xy' + x^2 y'' \; \Rightarrow \; y''' \, (0) \; = 2y \, (0) \; = \; 2;$$

$$y^{(iv)} \; = \; 6y' + 6xy'' + x^2 y''' \Rightarrow y^{(iv)} \, (0) = \; 6y' \, (0) \; = \; - \, 6,$$

etc. Hence, the Taylor's series in Eq. (3.2) for y about $x_0 = 0$ is

$$y = 1 + (- \, 1) \, x + 0.x^2 \, / 2! + 2x^3 / 3! + (- \, 6) \, x^4 / 4! + \ldots$$

$$= 1 - x + x^3 / 3 - x^4 / 4 + \ldots \, \infty. \tag{3.7}$$

Therefore, for the given values of $x = 0 \cdot 1$ and $0 \cdot 2$, we have:

$$y(0\cdot1) = 1 - \cdot1 + \cdot001/3 - \cdot0001/4$$

$$= 1\cdot000333 - \cdot100025 = \cdot900308 \sim 0\cdot90031;$$

and

$$y(0\cdot2) = 1 - \cdot2 + \cdot008/3 - \cdot0016/4$$

$$= 1\cdot00267 - \cdot20040 = \cdot80227. \; //$$

Example 3.2. Using Taylor's series method solve the differential equation

$$y' = y^2 + x, \qquad y(0) = 1; \tag{3.8}$$

and compute y at $x = 0\cdot1, 0\cdot2$.

Solution. For the given condition: $y(0) = 1$,

$$y' = y^2 + x \quad \Rightarrow \quad y'(0) = 1.$$

Successive derivatives of y' w.r.t. x are

$$y'' = 2y\,y' + 1 \; \Rightarrow \quad y''(0) = 3,$$

$$y''' = 2\,(y')^2 + 2y\,y'' \quad \Rightarrow \quad y'''(0) = 2 + 2\times3 = 8,$$

$$y^{(iv)} = 6y'\,y'' + 2y\,y''' \qquad \Rightarrow \quad y^{(iv)}(0) = 18 + 16 = 34, \; \text{etc.}$$

Therefore, the Taylor's series (3.2) for y about the point $x = 0$ becomes

$$y = 1 + x + 3x^2/2! + 8x^3/3! + 34\,x^4/4! + \dots$$

$$= 1 + x + 3x^2/2 + 4x^3/3 + 17\,x^4/12 + \dots$$

$$= 1 + x + (1.5)\,x^2 + (1.33333)\,x^3 + (1.41667)\,x^4 + \dots$$

Hence, we have

$$y(0\cdot1) = 1 + \cdot1 + \cdot01500 + \cdot00133 + \cdot00014 = 1\cdot11647;$$

and

$$y(0\cdot2) = 1 + \cdot2 + 1\cdot5\times\cdot04 + 1\cdot33333\times\cdot008 + 1\cdot41667\times\cdot0016$$

$$= 1 + \cdot2 + \cdot060 + \cdot01067 + \cdot00227 = 1\cdot27294. \; //$$

§ 4. Euler's method for the solution of Eq. (1.1)

Let P (x_0, y_0) be the initial point and Q any other point on the solution curve of Eq. (1.1). Let us divide the line of their abscissae, LM, into n sub-intervals each of very small length, say h, at L_1, L_2,... If the ordinate through L_1 meet the curve in $P_1(x_0+h, y_1)$ then

$$y_1 = L_1 P_1 = L_1 R_1 + R_1 P_1$$

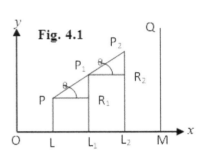

Fig. 4.1

$$= LP + (PR_1)\tan\theta = y_0 + h\,(dy/dx)_P,$$
$$\tag{4.1}$$

Similarly, if the ordinate through L_2 meet the curve in $P_2(x_0+2h, y_2)$ then

$$y_2 = L_2 P_2 = L_2 R_2 + R_2 P_2$$

$$= L_1 P_1 + (P_1 R_2)\tan\theta = y_1 + h\,(dy/dx)_{P_1} = y_1 + hf(x_0+h, y_1). \tag{4.2}$$

Continuing the process n times we derive

$$y_n = MQ = y_{n-1} + hf\{x_0 + (n-1)h, y_{n-1}\}, \tag{4.3}$$

that determines an appropriate solution of Eq. (1.1) by Euler's (or Euler-Cauchy) method.

Example 4.1. Using Euler's method compute an approximate value of y corresponding to $x = 1$ if

$$dy/dx = x + y, \qquad y(0) = 1. \tag{4.4}$$

Solution. Taking $n = 10$ and $h = 0.1$ (which is sufficiently small) various calculations are as follows:

x	y	$x + y = dy/dx$	preceding $y + h\,(dy/dx) = $ new y
0	1	$0 + 1 = 1$	$1 + \cdot 1 \times 1 = 1 \cdot 1$
$\cdot 1$	$1 \cdot 1$	$\cdot 1 + 1 \cdot 1 = 1 \cdot 2$	$1 \cdot 1 + \cdot 1 \times 1 \cdot 2 = 1 \cdot 1 + \cdot 12 = 1 \cdot 22$
$\cdot 2$	$1 \cdot 22$	$\cdot 2 + 1 \cdot 22 = 1 \cdot 42$	$1 \cdot 22 + \cdot 1 \times 1 \cdot 42 = 1 \cdot 22 + \cdot 14 = 1 \cdot 36$

·3	1·36	·3 + 1·36 = 1·66	1·36 + ·1×1·66 = 1·36 + ·17 = 1·53
·4	1·53	·4 + 1·53 = 1·93	1·53 + ·1×1·93 = 1·53 + ·19 = 1·72
·5	1·72	·5 + 1·72 = 2·22	1·72 + ·1×2·22 = 1·72 + ·22 = 1·94
·6	1·94	·6 + 1·94 = 2·54	1·94 + ·1×2·54 = 1·94 + ·25 = 2·19
·7	2·19	·7 + 2·19 = 2·89	2·19 + ·1×2·89 = 2·19 + ·29 = 2·48
·8	2·48	·8 + 2·48 = 3·28	2·48 + ·1×3·28 = 2·48 + ·33 = 2·81
·9	2·81	·9 + 2·81 = 3·71	2·81 + ·1×3·71 = 2·81 + ·37 = 3·18
1·0	3·18		

Thus, the appropriate value of y at $x = 1$ is $3·18$. //

§ 5. Improved Euler's method

A better approximation $y_1^{(1)}$ of $y_1 = y (x_0 + h)$ can be taken by considering the slope of the curve as the mean of the slopes of its tangents at P and P_1 (in Fig.4.1):

$$y_1^{(1)} = y_0 + (h / 2) \{f(x_0, y_0) + f(x_0 + h, y_1)\}, \qquad (5.1)$$

where y_1 is found by Euler's method vide equation (4.1).

Note 5.1. The Eq. (4.1) determines the *predictor* (value) while Eq. (5.1) the *corrector* (value) of y_1.

A further improved approximation $y_1^{(2)}$ of $y_1^{(1)}$ is similarly obtained as

$$y_1^{(2)} = y_0 + (h / 2)\{f(x_0, y_0) + f(x_0 + h, y_1^{(1)})\}. \qquad (5.2)$$

The process is repeated till two consecutive values of y become equal. Once y_1 is obtained to a desired degree of accuracy, y corresponding to the point L_2 is found from the predictor y_2 as given by Eq. (4.2), and a better approximation is obtained from the corrector:

$$y_2^{(1)} = y_1 + (h/2)\{f(x_0 + h, y_1) + f(x_0 + 2h, y_2)\}. \qquad (5.3)$$

Again this step is repeated until y_2 becomes stationary. Thereafter, we proceed to calculate y_3 on the similar lines.

Note 5.2. The improved Euler's method is also known as *predictor-corrector method.*

§ 6. Runge's method for the solution of Eq. (1.1)

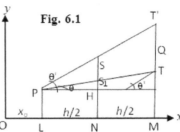

Fig. 6.1

The slope of the curve through the point $P_0(x_0, y_0)$ is $f(x_0, y_0)$. Integrating (1.1) from (x_0, y_0) to $(x_0 + h, y_0 + k)$ we derive

$$k \equiv \int_{y_0}^{y_0 + k} dy = \int_{x_0}^{x_0 + h} f(x, y) \, dx. \qquad (6.1)$$

Let N be the mid-point of $LM = h$. The values of x at the points L, N, M are $x_L = x_0$, $x_N = x_0 + h/2$, $x_M = x_0 + h$. The corresponding values of y at these points are $y_0 = LP$,

$$y_s = NS = NH + HS = LP + (PH) \tan \theta$$

$$= y_0 + (h/2)(dy/dx)_P = y_0 + (h/2) f(x_0, y_0). \qquad (6.2)$$

Also,

$$y_T = MT = MR + RT = LP + (PR) \tan \theta = y_0 + h f(x_0, y_0). \qquad (6.3)$$

The value of y at $Q(x_0 + h, y_Q)$ is given by the point T′, where the line drawn through P with the slope $T(x_0 + h, y_T)$ meets MQ. Slope of the curve at T, by Eq. (6.3), is

$$\tan \theta' = f(x_0 + h, y_T) = f\{x_0 + h, y_0 + h f(x_0, y_0)\}, \qquad (6.4)$$

Therefore,

$$y_Q \sim y_{T'} = MT' = MR + RT' = LP + (PR) \tan \theta',$$

or, by (6.4)

$$y_Q = y_0 + h f\{x_0 + h, y_0 + h f(x_0, y_0)\}. \qquad (6.5)$$

Further, the values of $f(x, y)$ at P, S and Q are

$$f_P = f(x_0, y_0), \quad f_S = f(x_0 + h/2, y_S), \quad f_Q = f(x_0 + h, y_Q). \qquad (6.6)$$

Hence, the integral in Eq. (6.1) reduces to

$$k = (h/6)\{f_P + 4f_S + f_Q\}, \tag{6.7a}$$

where Simpson's (one-third) rule [1] has been applied. Putting from Eq. (6.6), we thus have

$$k = (h/6)\{f(x_0, y_0) + 4f(x_0 + h/2, y_S) + f(x_0 + h, y_Q)\}, \tag{6.7b}$$

$$= (1/6)\{hf(x_0, y_0) + 4hf(x_0 + h/2, y_S) + hf(x_0 + h, y_Q)\}, \tag{6.7c}$$

determining a sufficiently accurate value of k and, therefore, that of $y = y_0 + k$.

6.1. Working rule: We compute

$$k_1 = hf_P = hf(x_0, y_0), \qquad k_2 = hf_S = hf(x_0 + h/2, y_0 + k_1/2), \tag{6.8}$$

$$k' = hf(x_0 + h, y_T) = hf(x_0 + h, y_0 + k_1),$$

$$k_3 = hf_Q = hf(x_0 + h, y_0 + k'). \tag{6.9}$$

Hence, Eq. (6.7c) finally reduces to

$$k = (1/6)(k_1 + 4k_2 + k_3). \tag{6.10}$$

Note 6.1. Above value of k is called the *weighted mean* of k_1, k_2, k_3.

Example 6.1. Given that there holds Eq. (4.4), find an approximate value of y at $x = 0 \cdot 2$ by Runge's method.

Solution. We have

$$x_0 = 0, \qquad y_0 = 1, \qquad h = 0 \cdot 2, \qquad f(x_0, y_0) = f(0,1) = 0 + 1 = 1.$$

---

[1] $\displaystyle \int_{x_0}^{x_0 + nh} f(x)\, dx$

$$= (h/3)\{(f_0 + f_n) + 4(f_1 + f_3 + \ldots + f_{n-1}) + 2(f_2 + f_4 + \ldots + f_{n-2})\}.$$

Presently, in Fig. 6.1,

$$nh = LM = 2(h/2), \quad f_0 = f_P, \quad f_n = f_Q, \quad f_1 = f_S, \quad f_2 = 0.$$

Therefore, Eqs. (6.8) - (6.10) determine

$$k_1 = \cdot2, \quad k_2 = \cdot2\, f(\cdot1,\, 1\cdot1) = \cdot2\, (\cdot1+1\cdot1) = \cdot2(1\cdot2) = \cdot24,$$

$$k' = \cdot2\, f(\cdot2,\, 1 + \cdot2) = \cdot2\, (\cdot2 + 1\cdot2) = \cdot2(1\cdot4) = \cdot28,$$

$$k_3 = \cdot2\, f(\cdot2,\, 1 + \cdot28) = \cdot2\, (\cdot2 + 1\cdot28) = \cdot2(1\cdot48) = \cdot296,$$

and

$$k = (1/6)\,(\cdot2 + \cdot96 + \cdot296) = (1\cdot456)/6 = \cdot2426.$$

Hence, the required approximate value of y is $y_0 + k = 1\cdot2426$. //

§ 7. Runge-Kutta method

Solution of differential equations by Taylor's series method involves the computation of higher order derivatives of the function. Runge-Kutta method does not require computation of such derivatives. It agrees with Taylor's series method up to the terms containing h^r, where r differs from method to method and is called the *order of method*.

Euler's method, improved Euler's method and Runge's method are indeed the Runge-Kutta methods of the *first, second* and *third* order respectively. The *fourth* order Runge-Kutta method is generally used and is often referred to as *Runge-Kutta method* only.

7.1. Working rule: To find the increment k in y corresponding to an increment h in x by Runge-Kutta method we compute the following quantities for the solution of differential equation (1.1) together with the condition (1.2):

$$k_1 = h f(x_0, y_0), \quad k_2 = h f(x_0 + h/2,\, y_0 + k_1/2), \qquad (7.1)$$

$$k_3 = h f(x_0 + h/2,\, y_0 + k_2/2), \qquad k_4 = h f(x_0 + h,\, y_0 + k_3);$$

and finally compute the weighted mean of above quantities:

$$k = (1/6)\,(k_1 + 2k_2 + 2k_3 + k_4). \qquad (7.2)$$

Therefore, y_1 is obtained:

$$y_1 = y_0 + k. \qquad (7.3)$$

Example 7.1. Evaluate an approximate value of y at $x = 0\cdot2$ by solving the equation (4.4) by Runge-Kutta method.

Solution. At $x = 0$, we have $x_0 = 0$, $y_0 = 1$, $h = \cdot 2$, $f(x_0, y_0) = x_0 + y_0$ $= 1$. Hence,

$$k_1 = \cdot 2, \quad k_2 = \cdot 2 f(0 + \cdot 1, 1 + \cdot 1) = \cdot 2\,(\cdot 1 + 1\cdot 1) = \cdot 2\,(1\cdot 2) = \cdot 24,$$

$$k_3 = \cdot 2 f(0 + \cdot 1, 1 + \cdot 12) = \cdot 2\,(\cdot 1 + 1\cdot 12) = \cdot 2\,(1\cdot 22) = \cdot 244,$$

$$k_4 = \cdot 2 f(0 + \cdot 2, 1 + \cdot 244) = \cdot 2\,(\cdot 2 + 1\cdot 244) = \cdot 2\,(1\cdot 444) = \cdot 2888.$$

Therefore, Eq. (7.2) determines

$$k = (\cdot 2 + \cdot 48 + \cdot 488 + \cdot 2888)/6 = (1\cdot 4568)/6 = \cdot 2428$$

$\Rightarrow$

$$(y)_{x = \cdot 2} = y_0 + k = 1\cdot 2428. \; //$$

Example 7.2. Evaluate an approximate value of y at $x = 0\cdot 2$ in steps of $0\cdot 1$ when there holds Eq. (3.8).

Solution. (i) At $x = 0$, we have

$$x_0 = 0, \quad y_0 = 1, \quad h = 0\cdot 1, \quad f(x_0, y_0) = f(0, 1) = (1)^2 + 0 = 1, \text{ by Eq. (3.8).}$$

Accordingly, Eq. (7.1) determines $k_1 = \cdot 1$,

$$k_2 = \cdot 1 f(0 + \cdot 05, 1 + \cdot 05) = \cdot 1\{(1\cdot 05)^2 + \cdot 05\} = \cdot 1\,(1\cdot 1525) = \cdot 11525,$$

$$k_3 = \cdot 1 f(0 + \cdot 05, 1 + \cdot 0576) = \cdot 1\{(1\cdot 0576)^2 + \cdot 05\}$$

$$= \cdot 1(1\cdot 1185 + \cdot 05) = \cdot 11685,$$

$$k_4 = \cdot 1 f(0 + \cdot 1, 1 + \cdot 11685) = \cdot 1\{(1\cdot 1169)^2 + \cdot 1\} = \cdot 1347$$

Hence, from Eq. (7.2) there follows

$$k = (\cdot 1 + \cdot 2305 + \cdot 2337 + \cdot 1347)/6 = (\cdot 6989)/6 = \cdot 1165;$$

$\Rightarrow$

$$(y)_{x = \cdot 1} = y_0 + k = 1\cdot 1165.$$

(ii) $\qquad x_1 = x_0 + h = 0 + \cdot 1 = \cdot 1, \qquad y_1 = 1\cdot 1165, \qquad h = 0\cdot 1,$

$$f(x_1, y_1) = f(\cdot 1, 1\cdot 1165) = (1\cdot 1165)^2 + \cdot 1 = 1\cdot 3466.$$

Accordingly, Eq. (7.1) determines

$$k_1 = h f(x_1, y_1) = \cdot 1 \, (1 \cdot 3466) = \cdot 1347,$$

$$k_2 = h f(x_1 + h/2, y_1 + k_1/2) = \cdot 1 f(\cdot 15, 1 \cdot 1165 + \cdot 0674)$$

$$= \cdot 1 f(\cdot 15, 1 \cdot 1839) = \cdot 1 \, \{(1 \cdot 1839)^2 + \cdot 15\} = \cdot 1552,$$

$$k_3 = h f(x_1 + h/2, y_1 + k_2/2) = \cdot 1 f(\cdot 15, 1 \cdot 1165 + \cdot 0776)$$

$$= \cdot 1 f(\cdot 15, 1 \cdot 1941) = \cdot 1 \, \{(1 \cdot 1941)^2 + \cdot 15\} = \cdot 1 \, (1 \cdot 5759) = \cdot 1576,$$

$$k_4 = h f(x_1 + h, y_1 + k_3) = \cdot 1 f(\cdot 1 + \cdot 1, 1 \cdot 1165 + \cdot 1576)$$

$$= \cdot 1 f(\cdot 2, 1 \cdot 2741) = \cdot 1 \, \{(1 \cdot 2741)^2 + \cdot 2\} = \cdot 1(1 \cdot 8233) = \cdot 1823..$$

Hence, from Eq. (7.2) there follows

$$k = (\cdot 1347 + \cdot 3104 + \cdot 3152 + \cdot 1823)/6 = (\cdot 9426)/6 = \cdot 1571;$$

$$\Rightarrow \quad (y)_{x = \cdot 2} = y_1 + k = 1 \cdot 1165 + \cdot 1571 = 1 \cdot 2736. \, //$$

§ 8. Milne's method

Given the differential equation (1.1) together with the initial condition (1.2) we compute an approximate value of y for $x = x_0 + nh$ by Milne's method in the following.

Computing $y_1 = y(x_0 + h)$, $y_2 = y(x_0 + 2h)$, $y_3 = y(x_0 + 3h)$ by Picard's or Taylor's series method, we evaluate

$$f_0 = f(x_0, y_0), \quad f_1 = f(x_0 + h, y_1), \quad f_2 = f(x_0 + 2h, y_2),$$

$$f_3 = f(x_0 + 3h, y_3). \tag{8.1}$$

To find $y_4 = y(x_0 + 4h)$, we use

$$y_4 = y_0 + \int_{x_0}^{x_0 + 4h} f(x, y) \, dx, \tag{8.2}$$

and substitute for $f(x, y)$ from *Newton's forward interpolation formula*:

$$f(x, y) = f_0 + n \, \Delta f_0 + \{n \, (n - 1)/2\} \Delta^2 f_0$$

$$+ \{n(n-1)(n-2)/6\}\Delta^3 f_0 + \ldots \tag{8.3}$$

Putting $x = x_0 + nh$ so that $dx = h\,dn$, and from Eq. (8.3) in Eq. (8.2), we get

$$y_4 = y_0 + h\int_{n=0}^{4} \{f_0 + n\Delta f_0 + \{n(n-1)/2\}\Delta^2 f_0$$

$$+ \{n(n-1)(n-2)/6\}\Delta^3 f_0 + \ldots\}dn = y_0 +$$

$$h\int_{n=0}^{4} \{f_0 + n\Delta f_0 + \{(n^2-n)/2\}\Delta^2 f_0 + \{(n^3 - 3n^2 + 2n)/6\}\Delta^3 f_0 + \ldots\}dn$$

$$= y_0 + h\Big[\; nf_0 + (n^2/2)\,\Delta f_0 + \{(1/2)(n^3/3 - n^2/2)\}\Delta^2 f_0$$

$$+ (1/6)\{(n^4/4 - n^3 + n^2)\}\Delta^3 f_0 + \ldots \;\Big]_0^4$$

$$= y_0 + h\,[4f_0 + 8\,\Delta f_0 + (20/3)\,\Delta^2 f_0 + (8/3)\,\Delta^3 f_0 + \ldots]. \tag{8.4}$$

Expressing the variations Δf_0, $\Delta^2 f_0$, $\Delta^3 f_0$ in terms of functional values:

$$f_1 = f_0 + \Delta f_0, \qquad f_2 = f_1 + \Delta f_1 \quad \text{and} \quad f_3 = f_2 + \Delta f_2 \tag{8.5}$$

$\Rightarrow$

$$\Delta f_0 = f_1 - f_0, \qquad \Delta f_1 = f_2 - f_1 \quad \text{and} \quad \Delta f_2 = f_3 - f_2. \tag{8.6}$$

Operating the first of relations (8.6) by Δ, and putting from Eq. (8.6) only, we derive

$$\Delta^2 f_0 = \Delta f_1 - \Delta f_0 = (f_2 - f_1) - (f_1 - f_0) = f_2 - 2f_1 + f_0. \tag{8.7}$$

Operating Eq. (8.7) further by Δ, we get

$$\Delta^3 f_0 = \Delta f_2 - 2\Delta f_1 + \Delta f_0 = (f_3 - f_2) - 2(f_2 - f_1) + (f_1 - f_0), \text{ by Eq. (8.6)}$$

$$= f_3 - 3f_2 + 3f_1 - f_0. \tag{8.8}$$

Neglecting the terms containing powers of Δ higher than 3 in Eq. (8.4) and putting from Eqs. (8.6) - (8.8) we evaluate

$$y_4 = y_0 + h\,[\,4f_0 + 8(f_1 - f_0) + (20/3)(f_2 - 2f_1 + f_0)$$

$$+ (8/3) (f_3 - 3 f_2 + 3 f_1 - f_0)] = y_0 + (4h/3) (2 f_1 - f_2 + 2 f_3). \quad (8.9)$$

The first approximation to f_4 is thus given by

$$f_4 = f(x_0 + 4h, y_4). \quad (8.10)$$

Note 8.1. Above value of y_4 is called a *predictor*.

A better approximation to y_4 is found by Simpson's rule:

$$y_4 = y_2 + (h/3) (f_2 + 4 f_3 + f_4), \quad (8.11)$$

which is called *corrector*? Repeating this process of approximation till y_4 becomes stationary, we evaluate $y_5 = y(x_0 + 5h)$ from the predictor formula (8.9):

$$y_5 = y_1 + (4h/3) (2 f_2 - f_3 + 2 f_4). \quad (8.12)$$

Further, a better approximation to y_5 is calculated from the corrector formula (8.11):

$$y_5 = y_3 + (h/3) (f_3 + 4 f_4 + f_5). \quad (8.13)$$

Repeating this process till y_5 becomes stationary, we proceed to calculate y_6 as before.

Example 8.1. Apply Milne's method to find a solution of the differential equation

$$y' \equiv dy/dx = x - y^2 = f(x, y) \quad (8.14)$$

in the range $0 \le x \le 1$ for the boundary condition

$$y(0) = 0. \quad (8.15)$$

Solution. (i) Using Picard's method a value of y is given by Eq. (2.2):

$$y = 0 + \int_0^x f(x, y) \, dx = \int_0^x (x - y^2) \, dx.$$

Thus, the first approximation y_1 of y, given by Eq. (2.3), becomes

$$y_1 = 0 + \int_0^x (x - 0) \, dx = x^2/2. \quad (8.16)$$

Applying Eq. (2.4), the second approximation to y is obtained:

$$y_2 = 0 + \int_0^x f(x, y_1)\, dx = \int_0^x f(x, x^2/2)\, dx$$

$$= \int_0^x (x - x^4/4)\, dx = x^2/2 - x^5/20. \tag{8.17}$$

Similarly, the third approximation, given by Eq. (2.5), is

$$y_3 = 0 + \int_0^x f(x, y_2)\, dx = \int_0^x f(x, x^2/2 - x^5/20)\, dx$$

$$= \int_0^x \{(x - (x^2/2 - x^5/20)^2\}\, dx = \int_0^x \{x - x^4/4 + x^7/20 - x^{10}/400\}\, dx$$

$$= x^2/2 - x^5/20 + x^8/160 - x^{11}/4400. \tag{8.18}$$

(ii) Choosing $h = 0.2$ and applying Milne's method we evaluate the starting values

$x = x_0$ $= 0$	$y_0 = 0$	$f_0 = f(x_0, y_0) = f(0, 0) = 0$
$x = .2$	$y_1 = y(x_0 + h) = y(.2)$ $= (.2)^2/2 = .02$	$f_1 = f(x_0 + h, y_1) = f(.2, .02)$ $= .2 - (.02)^2 = .1996$
$x = .4$	$y_2 = y(x_0 + 2h) = y(.4)$ $= (.4)^2/2 - (.4)^5/20 = .0795$	$f_2 = f(x_0 + 2h, y_2) = f(.4, .0795)$ $= .4 - (.0795)^2 = .3937$
$x = .6$	$y_3 = y(x_0 + 3h)$ $= y(.6) = .1762$	$f_3 = f(x_0 + 3h, y_3) = f(.6, .1762)$ $= .6 - (.1762)^2 = .5689$

Next, applying the predictor formula (8.9), y_4 is evaluated:

$x = .8$	$y_4 = \{4(.2)/3\}(2 \times .1996 - .3937 + 2 \times .5689\} = .3049$	$f_4 = f(x_0 + 4h, y_4) = f(.8, .3049)$ $= .8 - (.3049)^2 = .7070$

Also, Eq. (8.11) determines the corrector:

$$y_4 = \cdot0795 + (\cdot2/3)\,(\cdot3937 + 2\cdot2756 + \cdot7070) = \cdot0795 + (\cdot2/3)\,(3\cdot3763)$$

$$= \cdot0795 + \cdot2(1\cdot1254) = \cdot3046. \qquad (8.19)$$

Accordingly,

$$f_4 = f(x_0 + 4h, y_4) = f(\cdot8, \cdot3046) = \cdot8 - (\cdot3046)^2 = \cdot7072.$$

Applying the corrector formula (8.11) again for this value of f_4, we evaluate y_4:

$$y_4 = \cdot0795 + (\cdot2/3)(\cdot3937 + 2\cdot2756 + \cdot7072) = \cdot3046,$$

which is the same as above in (8.19).

(iii) Next, applying the predictor formula (8.12), at $x = 1$, we obtain

$$y_5 = \cdot02 + (\cdot8/3)(\cdot7874 - \cdot5689 + 1\cdot4144) = \cdot02 + (\cdot8/3)\,(1\cdot6329)$$

$$= \cdot02 + \cdot8\,(\cdot5443) = \cdot02 + \cdot4354 = \cdot4554.$$

Therefore,

$$f_5 = f(x_0 + 5h, y_5) = f(1, \cdot4554) = 1 - (\cdot4554)^2 = \cdot7926.$$

Also, the corrector formula (8.13) determines

$$y_5 = \cdot1762 + (\cdot2/3)\,(\cdot5689 + 2\cdot8288 + \cdot7926)$$

$$= \cdot1762 + (\cdot2/3)\,(4\cdot1903) = \cdot1762 + \cdot2\,(1\cdot3967) = \cdot4555;$$

and, therefore,

$$f_5 = f(x_0 + 5h, y_5) = f(1, \cdot4555) = 1 - (\cdot4555)^2 = \cdot7925.$$

Applying the corrector formula (8.13) again for above value of f_5, we get

$$y_5 = \cdot1762 + (\cdot2/3)\,(\cdot5689 + 2\cdot8288 + \cdot7925) = \cdot4555;$$

as before. Hence, y at $x = 1$ is $y_5 = \cdot4555$. //

§ 9. Adams-Bashforth method

Let the differential equation be given by Eq. (1.1) together with the initial condition vide Eq. (1.2). We compute the values of y by any method such as Taylor' series, Euler or Runge-Kutta method:

$$y_{-1} = y(x_0 - h), \qquad y_{-2} = y(x_0 - 2h), \qquad y_{-3} = y(x_0 - 3h).$$

Accordingly, the values of the function $f(x, y)$ are evaluated:

$$f_{-1} = f(x_0 - h, y_{-1}), \quad f_{-2} = f(x_0 - 2h, y_{-2}), \quad f_{-3} = f(x_0 - 3h, y_{-3}).$$

Substituting from *Newton's backward interpolation formula* for $f(x, y)$:

$$f(x, y) = f_0 + n \, \Delta f_0 + \{ n(n+1)/2 \} . \, \Delta^2 f_0$$

$$+ \{ n(n+1)(n+2)/6 \} . \, \Delta^3 f_0 + \dots, \tag{9.1}$$

the Eq. (2.2) determines the first approximation y_1:

$$y_1 = y_0 + \int_{x_0}^{x_0+h} f(x, y_0) \, dx \tag{9.2}$$

$$= y_0 + \int_{x_0}^{x_0+h} [f_0 + n\Delta f_0 + \{n(n+1)/2\}\Delta^2 f_0$$

$$+ \{n(n+1)(n+2)/6\}\Delta^3 f_0 + \dots] \, dx.$$

Putting $x = x_0 + nh$ so that $dx = h \, dn$, above relation simplifies to

$$y_1 = y_0 + h \int_{n=0}^{1} [f_0 + n \, \Delta f_0 + \{n(n+1)/2\} . \, \Delta^2 f_0$$

$$+ \{ n(n+1)(n+2)/6 \} . \, \Delta^3 f_0 + \dots] . \, dn$$

$$= y_0 + h \left[n f_0 + (n^2/2) . \Delta f_0 + (n^3/6 + n^2/4) . \Delta^2 f_0 \right.$$

$$\left. + (n^4/24 + n^3/6 + n^2/6) . \Delta^3 f_0 + \dots \right]_0^1$$

$$= y_0 + h [f_0 + (1/2) \Delta f_0 + (5/12) \Delta^2 f_0 + (3/8) \Delta^3 f_0 + \dots] . \tag{9.3}$$

Neglecting the terms involving order of differentiation higher than three and changing the variations Δf_0, $\Delta^2 f_0$, $\Delta^3 f_0$ into the functionals f_0, f_1, f_2, f_3:

$$f_{-1} = f_0 - \Delta f_0, \qquad f_{-2} = f_{-1} - \Delta f_{-1}, \qquad f_{-3} = f_{-2} - \Delta f_{-2}$$

$$\Rightarrow$$

$$\Delta f_0 = f_0 - f_{-1}, \qquad \Delta f_{-1} = f_{-1} - f_{-2}, \qquad \Delta f_{-2} = f_{-2} - f_{-3}.$$

Their successive operations by Δ also yield

$$\Delta^2 f_0 = \Delta f_0 - \Delta f_{-1} = (f_0 - f_{-1}) - (f_{-1} - f_{-2}) = f_0 - 2f_{-1} + f_{-2},$$

$$\Delta^3 f_0 = \Delta f_0 - 2\Delta f_{-1} + \Delta f_{-2} = (f_0 - f_{-1}) - 2(f_{-1} - f_{-2}) + (f_{-2} - f_{-3})$$

$$= f_0 - 3f_{-1} + 3f_{-2} - f_{-3}.$$

Consequently, Eq. (9.3) reduces to

$$y_1 = y_0 + h\,[\,f_0 + (1/2)\,(f_0 - f_{-1}) + (5/12)\,(f_0 - 2f_{-1} + f_{-2})$$

$$+ (3/8)\,(f_0 - 3f_{-1} + 3f_{-2} - f_{-3})]$$

$$= y_0 + (h/24)\,(55\,f_0 - 59\,f_{-1} + 37\,f_{-2} - 9\,f_{-3}). \tag{9.4}$$

This formula is called *Adams-Bashforth predictor formula.*

We now evaluate the corresponding value of the function f for above approximation:

$$f_1 = f(x_0 + h, y_1). \tag{9.5}$$

To find a better approximation to y, we derive a corrector formula by substituting for $f(x, y)$ in Eq. (9.2) from the Newton's backward interpolation formula at f_1:

$$f(x, y) = f_1 + n\,\Delta f_1 + \{n\,(n+1)/2\}\Delta^2 f_1 + \{n\,(n+1)(n+2)/6\}\Delta^3 f_1 + \dots$$

Hence, Eq. (9.2) gives

$$y_1 = y_0 + \int_{x_0}^{x_1} [\,f_1 + n\,\Delta f_1 + \{n\,(n+1)/2\}\Delta^2 f_1$$

$$+ \{n\,(n+1)(n+2)/6\}\Delta^3 f_1 + \dots]\,dx.$$

Putting $x = x_1 + nh$ so that $dx = h.dn$ and using $x_0 - x_1 = -h$, above relation further reduces to

$$y_1 = y_0 + h \int_{n=-1}^{0} [f_1 + n \, \Delta f_1 + \{n \, (n+1)/2\} \Delta^2 f_1$$

$$+ \{n \, (n+1) \, (n+2)/6\} \Delta^3 f_1 + \ldots] \, dn$$

$$= y_0 + h \Big[\, n f_1 + (n^2/2) \, \Delta f_1 + (n^3/6 + n^2/4) \, \Delta^2 f_1$$

$$+ (n^4/24 + n^3/6 + n^2/6) \, \Delta^3 f_1 + \ldots \ \Big]_{-1}^{0}$$

$$= y_0 + h \, [f_1 - (1/2) \Delta f_1 - (1/12) \, \Delta^2 f_1 - (1/24) \, \Delta^3 f_1 + \ldots] \qquad (9.6)$$

Neglecting the terms involving order of differentiation higher than three and expressing the variations Δf_1, $\Delta^2 f_1$, $\Delta^3 f_1$ in terms of the functional values f_0, f_1, f_{-1}, f_{-2}:

$$f_0 = f_1 - \Delta f_1, \qquad\qquad f_{-1} = f_0 - \Delta f_0, \ \ f_{-2} = f_{-1} - \Delta f_{-1}$$

$$\Rightarrow$$

$$\Delta f_1 = f_1 - f_0, \qquad \Delta f_0 = f_0 - f_{-1}, \ \ \Delta f_{-1} = f_{-1} - f_{-2}.$$

Their successive operations by Δ also yield

$$\Delta^2 f_1 = \Delta f_1 - \Delta f_0 = (f_1 - f_0) - (f_0 - f_{-1}) = f_1 - 2 f_0 + f_{-1},$$

$$\Delta^3 f_1 = \Delta f_1 - 2\Delta f_0 + \Delta f_{-1}$$

$$= (f_1 - f_0) - 2 \, (f_0 - f_{-1}) + (f_{-1} - f_{-2}) = f_1 - 3 f_0 + 3 f_{-1} - f_{-2}.$$

Consequently, Eq. (9.6) reduces to

$$y_1 = y_0 + h \, [f_1 - (f_1 - f_0)/2 - (f_1 - 2 f_0 + f_{-1})/12$$

$$- (f_1 - 3f_0 + 3f_{-1} - f_{-2})/24] = y_0 + (h/24) \, (9f_1 + 19f_0 - 5f_{-1} + f_{-2}), \qquad (9.7)$$

giving a *corrector formula*. The function f, for above value of y_1, is evaluated by Eq. (9.5). Repeating the process till y_1 becomes stationary. Next, we proceed to compute y_2 as above.

Example 9.1. Given the differential equation

$$dy \, / \, dx = x^2 \, (1 + y), \qquad\qquad (9.8)$$

together with the boundary conditions

$$y(1) = 1, \ y(1\cdot1) = 1\cdot233, \ y(1\cdot2) = 1\cdot548 \text{ and } y(1\cdot3) = 1\cdot979,$$

evaluate $y(1\cdot4)$ by Adams-Bashforth method.

Solution. Given that $f(x, y) = x^2(1 + y)$ and $h = \cdot1$ the starting values of the Adams-Bashforth method are

$x = 1$	$y_{-3} = y(1) = 1$	$f_{-3} = (1)^2(1+1) = 2$
$x = 1\cdot1$	$y_{-2} = y(1\cdot1) = 1\cdot233$	$f_{-2} = (1\cdot1)^2(1 + 1\cdot233) = 2\cdot702$
$x = 1\cdot2$	$y_{-1} = y(1\cdot2) = 1\cdot548$	$f_{-1} = (1\cdot2)^2(1 + 1\cdot548) = 3\cdot669$
$x = 1\cdot3$	$y_0 = y(1\cdot3) = 1\cdot979$	$f_0 = (1\cdot3)^2(1 + 1\cdot979) = 5\cdot035$

Applying the predictor formula (9.4), the value of y_1 is evaluated:

$$y_1 = 1\cdot979 + (\cdot1/24)(55 \times 5\cdot035 - 59 \times 3\cdot669 + 37 \times 2\cdot702 - 9 \times 2)$$

$$= 1\cdot979 + (\cdot1/24)(142\cdot428) = 1\cdot979 + \cdot5935 = 2\cdot573.$$

Accordingly,

$$f_1 = f(1\cdot4, 2\cdot573) = (1\cdot4)^2(1 + 2\cdot573) = 7\cdot003.$$

Thereafter, applying the corrector formula (9.7), we get

$$y_1 = 1\cdot979 + (\cdot1/24)(9 \times 7\cdot033 + 19 \times 5\cdot035 - 5 \times 3\cdot669 + 2\cdot702)$$

$$= 1\cdot979 + (\cdot1/24)(143\cdot049) = 1\cdot979 + \cdot596 = 2\cdot575.$$

Thus, $y(1\cdot4) = y_1 = 2\cdot575.$ //

CHAPTER 5

CURVE FITTING

§ 1. Introduction

Having obtained some data from observations it is desired to express it in the form of a relation connecting the (two) variables involved therein. The relation so obtained is called an *empirical law*. There may be several equations of different types expressing the given data approximately. But, we find the equation of the curve of "best fit" that can be most suitable for predicting the unknown values. The process of finding this equation of "best fit" is called the *curve fitting*.

Given n pairs of observed values the given data can be fitted in an equation containing n arbitrary constants as n simultaneous equations are solvable for n unknowns. To fit above data in an equation containing less than n arbitrary constants the following methods are in use: (*i*) graphical method, (*ii*) method of moments, and (*iii*) method of least squares (given by Carl Friedrich Gauss).

The first of these methods does not give the values of unknowns as accurate as the other methods. The last method is the best to fit a unique curve to a given data. It is most commonly used and can be easily implemented on a computer.

§ 2. Method of least squares

Let the curve
$$y = a + bx + cx^2 \qquad (2.1)$$

is to be fitted to a given data (x_1, y_1), (x_2, y_2), ... , (x_n, y_n). Thus, for any x_i, $i = 1, 2, \ldots, n$, the observed value is y_i and let η_i be the expected value:
$$\eta_i = a + b x_i + c x_i^2. \qquad (2.2)$$

Therefore, the error is
$$E_i = y_i - \eta_i = y_i - a - b x_i - c x_i^2. \qquad (2.3)$$

The sum of squares of these errors is

$$S \equiv \sum_{i=1}^{n} (E_i)^2 = \sum_{i=1}^{n} (y_i - a - b x_i - c x_i^2)^2. \qquad (2.4)$$

For minimum S there should hold the conditions:

$$\partial S / \partial a \equiv -2 \Sigma (y_i - a - b x_i - c x_i^2) = 0$$

$$\Rightarrow \quad \sum_{i=1}^{n} y_i = \sum_{i=1}^{n} a + b \Sigma x_i + c \Sigma x_i^2 = na + b \Sigma x_i + c \Sigma x_i^2; \quad (2.5)$$

$$\partial S / \partial b \equiv -2 \Sigma x_i (y_i - a - b x_i - c x_i^2) = 0$$

$$\Rightarrow$$

$$\Sigma x_i y_i = a \Sigma x_i + b \Sigma x_i^2 + c \Sigma x_i^3; \qquad (2.6)$$

and

$$\partial S / \partial c \equiv -2 \Sigma x_i^2 (y_i - a - b x_i - c x_i^2) = 0$$

$$\Rightarrow$$

$$\Sigma x_i^2 y_i = a \Sigma x_i^2 + b \Sigma x_i^3 + c \Sigma x_i^4. \qquad (2.7)$$

The equations (2.5) – (2.7) being three simultaneous linear equations in (three) unknowns a, b, c are uniquely solvable. The values of these constants when substituted in (2.1) determine the desired curve of best fit.

Note 2.1. The Eqs. (2.5) - (2.7) are called the *normal equations*.

Example 2.1. Fit the straight line

$$y = a + bx \qquad (2.8)$$

to a data (x_i, y_i), $i = 1, 2, \ldots, n$.

Solution. Comparing (2.8) and (2.1) we note that $c = 0$ in the present case. Hence, the normal equations (2.5) and (2.6), for (2.8), reduce to

$$n a + b \Sigma x_i = \Sigma y_i; \qquad (2.5a)$$

and

$$a\Sigma x_i + b \Sigma x_i^2 = \Sigma x_i y_i. \qquad (2.6a)$$

respectively. Solving these simultaneous equations in a and b, we derive

$$a = \{(\Sigma x_i^2) (\Sigma y_i) - (\Sigma x_i) (\Sigma x_i y_i)\} / \{n \Sigma x_i^2 - (\Sigma x_i)^2\}, \qquad (2.9)$$

and

$$b = \{n \Sigma x_i y_i - (\Sigma x_i) (\Sigma y_i)\} / \{n \Sigma x_i^2 - (\Sigma x_i)^2\}. \qquad (2.10)$$

Substituting for the values of a, b in Eq. (2.8) we get the best fit for above data. //

Example 2.2. (i) Let the pull P be required to lift a load W kg. by means of a pulley block. Find a linear relation of the form

$$P = a + bW \qquad (2.11)$$

connecting P and W by using the following data: $P = 12, 15, 21, 25$; $W = 50, 70, 100, 120$.

(ii) Compute P when $W = 150$ kg. weight.

Solution. (i) The normal equations (2.5a) and (2.6a) for the present data are

$$4a + b \Sigma W = \Sigma P, \qquad (2.12)$$

and

$$a \Sigma W + b \Sigma W^2 = \Sigma WP. \qquad (2.13)$$

We compute the following values:

W	P	W^2	WP
50	12	2,500	600
70	15	4,900	1,050
100	21	10,000	2,100
120	25	14,400	3,000
$\Sigma W = 340$	$\Sigma P = 73$	$\Sigma W^2 = 31,800$	$\Sigma WP = 6,750$

Therefore, the equations (2.12) and (2.13) reduce to

$$4a + 340 b = 73 \qquad \Rightarrow \qquad 2a + 170 b = 36 \cdot 5,$$

and

$$340 a + 31800 b = 6750 \qquad \Rightarrow \qquad 34 a + 3180 b = 675.$$

Solving these simultaneous equations, we obtain

$$b = (675 - 17 \times 36 \cdot 5) / (3180 - 17 \times 170) = \cdot 1879,$$

and

$$a = (73 - 340 \times \cdot 1879) / 4 = 2 \cdot 2785;$$

for which Eq. (2.11) reduces to

$$P = 2 \cdot 2785 + (\cdot 1879) \, W. \tag{2.11a}$$

(ii) Putting $W = 150$ in (2.11a), we compute $P = 30 \cdot 4635$ kg. wt. //

§ 3. Spline fittings

Piecewise approximation of a function by some polynomial is called *spline* approximation. Let a function $f(x)$ be defined on an interval $a \leq x \leq b$. Let the interval be partitioned into n subintervals with common end points called *nodes*:

$$a = x_0 < x_1 < x_2 < \dots < x_n = b, \tag{3.1}$$

and $g(x)$ be a function given by some polynomial in these intervals such that:

(*i*) there exists only one polynomial in every subinterval,
and
(*ii*) $g(x)$ is several times differentiable at the end points.

Thus, instead of approximating $f(x)$ by a single function in the entire interval $[a, b]$ we approximate $f(x)$ by n polynomials. This way we obtain approximating functions $g(x)$ which suit more in problems of approximation and interpolation. Such functions $g(x)$ are called *splines*. Such a nomenclature has a base in engineering problems involving thin rods. Splines are of much use in practical problems and their study is relatively newer.

The linear functions provide the simplest continuous piecewise polynomial approximations. However, such a function is not differentiable at the end points of above subintervals. Therefore, it is preferable to take functions possessing certain order of differentiation everywhere on the interval $[a, b]$.

We consider the cubic splines which are continuous functions possessing continuous derivatives up to the second order throughout the interval $[a, b]$ and its every sub-interval (x_{i-1}, x_i), $i = 1, 2, \dots, n$. The degree of these splines does not exceed 3. Thus, $g(x)$ consists of cubic polynomials one in each subinterval:

$$g(x_0) = f(x_0), \ g(x_1) = f(x_1), \ \dots, \ g(x_n) = f(x_n). \tag{3.2}$$

If, in addition, there also hold

$$g'(x_0) = k_0, \qquad\qquad g'(x_n) = k_n. \tag{3.3}$$

for some given numbers k_0 and k_n then the cubic splines are uniquely determined.

Example 3.1. Approximate $f(x) = x^4$ on the interval $-1 \le x \le 1$ by cubic splines $g(x)$ corresponding to the partition

$$-1 = x_0, \quad x_1 = 0, \quad x_2 = 1, \tag{3.4}$$

and satisfying Eq. (3.2):

$$\left.\begin{aligned}
g(x_0) &= g(-1) = f(x_0) = f(-1) = (-1)^4 = 1, \\
g(x_1) &= g(0) = f(0) = (0)^4 = 0, \\
g(x_2) &= g(1) = f(1) = (1)^4 = 1;
\end{aligned}\right\} \tag{3.5}$$

and

$$g'(-1) = f'(-1) = 4(-1)^3 = -4, \ g'(1) = f'(1) = 4(1)^3 = 4. \tag{3.6}$$

Solution. Let the cubic splines $g(x)$ in the subintervals $(-1, 0)$ and $(0, 1)$ be defined as

$$g_0(x) \equiv a_0 + a_1 x + a_2 x^2 + a_3 x^3, \qquad -1 \le x \le 0; \tag{3.7}$$

and

$$g_1(x) \equiv b_0 + b_1 x + b_2 x^2 + b_3 x^3, \qquad 0 \le x \le 1; \tag{3.8}$$

respectively. Evaluating their values as per Eq. (3.5) we get

$$g_0(-1) = a_0 - a_1 + a_2 - a_3 = f(-1) = 1, \quad g_0(0) = a_0 = f(0) = 0,$$

$$g_1(0) = b_0 = f(0) = 0, \qquad g_1(1) = b_0 + b_1 + b_2 + b_3 = f(1) = 1;$$

i.e.

$$a_0 = b_0 = 0, \qquad\qquad a_1 - a_2 + a_3 = -1, \tag{3.9}$$

and

$$b_1 + b_2 + b_3 = 1. \tag{3.10}$$

Also, differentiating the spline functions in Eqs. (3.7) and (3.8) with respect to x and putting from Eq. (3.6) we obtain

$$g'_o(-1) = a_1 + 2a_2(-1) + 3a_3(-1)^2 = a_1 - 2a_2 + 3a_3 = -4, \tag{3.11}$$
and
$$g'_1(1) = b_1 + 2b_2 + 3b_3 = 4. \tag{3.12}$$

Further, using

$$\left.\begin{array}{l} g'_o(0) = a_1 = g'_1(0) = b_1 \quad \Rightarrow \quad a_1 = b_1 \\[2mm] g''_o(0) = 2a_2 = g''_1(0) = 2b_2 \quad \Rightarrow \quad a_2 = b_2 \, . \end{array}\right\} \tag{3.13}$$

Hence, the relations (3.10) and (3.12) reduce to

$$a_1 + a_2 + b_3 = 1 \quad (3.10\text{a}); \qquad \text{and} \qquad a_1 + 2a_2 + 3b_3 = 4. \tag{3.12a}$$

Solving Eqs. (3.9), (3.11), (3.10a) and (3.12a), we find

$$a_1 = b_1 = 0, \quad a_2 = b_2 = -1, \ a_3 = -2, \ b_3 = 2.$$

Hence, Eqs. (3.7) and (3.8) assume the forms

$$g_o(x) = -x^2 - 2x^3, \ -1 \le x \le 0; \quad \text{and} \quad g_1(x) = -x^2 + 2x^3, \quad 0 \le x \le 1. \ //$$

Example 3.2. Show that for a given partition in Eq. (3.1) there exist $n+1$ unique cubic splines $g_o(x), g_1(x), \ldots, g_n(x)$ such that

$$g'_i(a) = g'_i(b) = 0; \tag{3.14}$$
and
$$g_i(x_j) = \delta_{ij} = 0 \ (\text{if } i \ne j), \quad 1 \ (\text{if } i = j). \tag{3.15}$$

CHAPTER 6

NUMERICAL INTEGRATON

§ 1. Trapezoidal rule for area of a region bounded by a plane curve

Let a plane curve C enclose certain Region R. Consider the largest segment AB of the region. We divide the segment into n equal parts each of width, say, h. Taking x-axis along the segment AB, we draw the perpendicular transversals of lengths $y_0, y_1, y_2, \ldots, y_{n+1}$ through these points. Let $P_0, P_1, P_2, \ldots, P_{n+1}$ be their

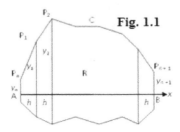

Fig. 1.1

end points lying on one side of the curve. Joining the end points of every two successive transversals on either side of the segment AB, the whole region R can be viewed as a totality of n trapeziums each of width h. Thus, the areas of these trapeziums are

$$h\,(y_0 + y_1)\,/\,2,\ \ h\,(y_1 + y_2)\,/\,2,\ \ldots,\ h\,(y_n + y_{n+1})\,/\,2.$$

Their sum determines an approximate area of the whole region:

$$A_T = (h/2).\ \{\,y_0 + 2(y_1 + y_2 + \ldots + y_n) + y_{n+1}\,\}$$

$$= (h/2).\ \{\,(y_0 + y_{n+1}) + 2(y_1 + y_2 + \ldots + y_n)\,\}. \tag{1.1}$$

Note 1.1. Above formula is called the *trapezoidal rule* for an approximate area of the region.

§ 2. Simpson's rule for an approximate area of a region

Let C be a plane curve lying between two parallels $x = a$ and $x = b$. The area enclosed by C, these two parallels and x-axis can be found if the curve is represented by some equation, say, $y = f(x)$, by the formula

$$A = \int_{x=a}^{b} f(x)\ dx. \tag{2.1}$$

On the other hand, if C does not possess any equation and only the coordinates of the points on it are known a method to find above area is demonstrated here. Let the points $P_1, P_2, \ldots, P_{2n+1}$ be taken on C so that their ordinates $y_1, y_2, \ldots, y_{2n+1}$ divide the whole region into $2n$ strips each of width, say, h. Thus, abscissae of these points are

$$a, \quad a+h, \quad a+2h, \quad \ldots, \quad a+2nh = b.$$

We consider a curve (which is usually a parabola) through the points P_1, P_2, P_3 providing the best possible approximation to C. Let its equation referred to the foot of the ordinate through P_2 as (new) origin be

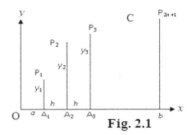

Fig. 2.1

$$y = \alpha + \beta x + \gamma x^2. \qquad (2.2)$$

Thus, the new coordinates of the points P_1, P_2, P_3 are $(-h, y_1)$, $(0, y_2)$ and $(-h, y_3)$ which satisfy the Eq. (2.2):

$$y_1 = \alpha - \beta h + \gamma h^2, \qquad (2.3)$$

$$y_2 = \alpha \qquad (2.4); \qquad y_3 = \alpha + \beta h + \gamma h^2. \qquad (2.5)$$

Adding Eqs. (2.3), (2.5), and putting from Eq. (2.4), we determine

$$y_1 - 2 y_2 + y_3 = 2 \gamma h^2. \qquad (2.6)$$

Next, putting from Eqs. (2.4) and (2.6) in (2.5), we evaluate:

$$\beta h = y_3 - y_2 - (y_1 - 2y_2 + y_3)/2 = (y_3 - y_1)/2. \qquad (2.7)$$

Therefore, the area bounded by the portion $P_1 P_2 P_3$ (of the parabolic path), x-axis and the ordinates of P_1, P_3 as per formula (2.1) is

$$\int_{x=-h}^{h} y \, dx = \int_{x=-h}^{h} (\alpha + \beta x + \gamma x^2) \, dx = 2 \int_{x=0}^{h} (\alpha + \gamma x^2) \, dx$$

$$= 2h. \, (\alpha + \gamma h^2 / 3) = 2h \{ y_2 + (y_1 - 2y_2 + y_3)/6 \}$$

$$= h (y_1 + 4y_2 + y_3)/3, \qquad (2.8)$$

where substitutions for α, γ are made from Eqs. (2.4) and (2.6) and Eq. (1.1.7) is applied. Similarly, the area bounded by the portion $P_3 P_4 P_5$ of some parabolic path, x-axis and the ordinates of points P_3, P_5 can be found as $h\,(y_3 + 4y_4 + y_5)\,/\,3$.

Also, the area bounded by the portion of the curve through the end points of the last two strips: P_{2n-1}, P_{2n} and P_{2n+1} shall be

$$h\,(y_{2n-1} + 4y_{2n} + y_{2n+1})\,/\,3.$$

Addition of all such areas determines an approximate area bounded by the entire curve C, x-axis and the ordinates $x = a$ and $x = b$:

$$A_S = (h/3)\{(y_1 + 4y_2 + y_3) + (y_3 + 4y_4 + y_5) + \ldots + (y_{2n-1} + 4y_{2n} + y_{2n+1})\}$$

$$= (h/3)\{y_1 + 4(y_2 + y_4 + y_6 + \ldots + y_{2n})$$

$$+ 2(y_3 + y_5 + y_7 + \ldots + y_{2n-1}) + y_{2n+1}\}. \tag{2.9}$$

Note 2.2. Above formula is called *Simpson's rule* (or more precisely, *Simpson's one-third rule*) for an approximate area of the region.

§ 3. Simpson's *three-eighth* rule for an approximate area of a region

Let C be a plane curve having equation

$$y = \alpha + \beta x + \gamma x^2 + \delta x^3. \tag{3.1}$$

Fig. 3.1

The area enclosed by the curve, x-axis and two parallels $x = 0$ and $x = 3h$ of y-axis can also be found by an alternate formula:

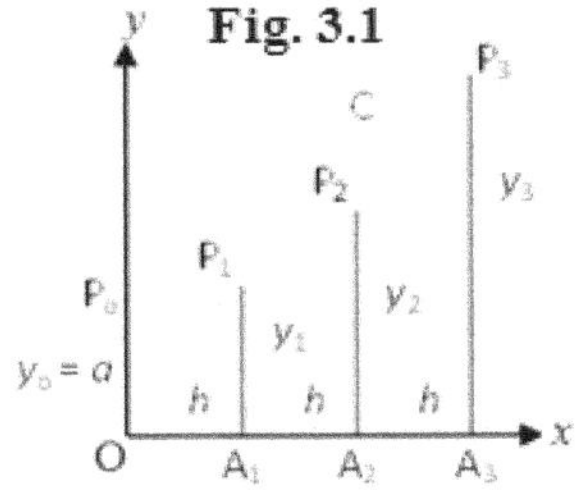

$$A_{S*} = (3h/8)\,(y_0 + 3y_1 + 3y_2 + y_3), \tag{3.2}$$

where y_r is the value of y at $x = rh$. To establish the result, we consider four points $P_o\,(0, y_0)$, $P_1\,(h, y_1)$, $P_2\,(2h, y_2)$ and $P_3\,(3h, y_3)$ on the curve. So, there hold the relations

$$y_0 = \alpha \tag{3.3}; \qquad y_1 = \alpha + \beta h + \gamma h^2 + \delta h^3, \tag{3.4}$$

$$y_2 = \alpha + 2\beta h + 4\gamma h^2 + 8\delta h^3, \tag{3.5}$$

$$y_3 = \alpha + 3\beta h + 9\gamma h^2 + 27\delta h^3. \tag{3.6}$$

Subtracting the sum of Eqs. (3.4) and (3.5) from the sum of Eqs. (3.3) and (3.6) we eliminate the coefficients α, β:

$$y_0 + y_3 - y_1 - y_2 = 4\gamma h^2 + 18\,\delta h^3 . \tag{3.7}$$

Also, subtraction of second multiple of Eq. (3.5) from the sum of Eqs. (3.4) and (3.6) yields another simultaneous relation in γ, δ:

$$y_1 + y_3 - 2y_2 = 2\gamma\, h^2 + 12\delta\, h^3 . \tag{3.8}$$

The last two, relations determine the coefficients γ, δ:

$$\delta h^3 = (-y_0 + 3y_1 - 3y_2 + y_3)\,/6, \quad \gamma h^2 = (2y_0 - 5y_1 + 4y_2 - y_3)\,/2 . \tag{3.9}$$

Putting from Eqs. (3.3) and (3.9) in (3.4) the coefficient β is, thus, determined:

$$\beta h = y_1 - y_0 - (2y_0 - 5y_1 + 4y_2 - y_3)\,/\,2 + (y_0 - 3y_1 + 3y_2 - y_3)\,/\,6$$

$$= (-11y_0 + 18y_1 - 9y_2 + 2y_3)\,/\,6. \tag{3.10}$$

Thus, the desired area as per formula (2.1) becomes

$$A_{S*} = \int_{x=0}^{3h} y\, dx = \int_{x=0}^{3h} (\alpha + \beta x + \gamma x^2 + \delta x^3)\, dx$$

$$= 3h\,(\,\alpha + 3\beta h\,/2 + 3\gamma h^2 + 27\delta h^3\,/4\,),$$

which, for Eqs. (3.3), (3.9) and (3.10) simplifies to

$$A_{S*} = 3h\,\{\,y_0 + (-11y_0 + 18y_1 - 9y_2 + 2y_3)\,/4$$

$$+ 3\,(2y_0 - 5y_1 + 4y_2 - y_3)\,/\,2 \ + 9\,(-y_0 + 3y_1 - 3y_2 + y_3)\,/\,8\,\}.$$

Collecting the coefficients of like terms above result assumes the form of Eq. (3.2). //

CHAPTER 7

INTEGRAL TRANSFORMS

§ 1. Integral transform

Let $K(s, t)$ be some function of variables s and t and let the integral

$$\int_{-\infty}^{\infty} K(s, t) F(t)\, dt \tag{1.1}$$

be convergent. Above integral is called the *Integral transform* of the function $F(t)$ and is denoted by $T\{F(t)\}$ or $f(s)$:

$$T\{F(t)\} \equiv f(s) = \int_{t=-\infty}^{\infty} K(s, t) F(t)\, dt. \tag{1.2}$$

The function $K(s, t)$ is called the *kernel* of the transformation. The variable s is a parameter independent of t. Choosing

$$K(s, t) = 0 \ \text{(when } t < 0), \ e^{-st} \ \text{(when } t \geq 0), \tag{1.3}$$

the integral transform (1.2) becomes

$$T\{F(t)\} = \int_{t=-\infty}^{0-\varepsilon} 0.\, F(t)dt + \int_{t=0}^{\infty} e^{-st}.F(t)\, dt$$

$$= \int_{t=0}^{\infty} e^{-st}.F(t)\, dt \equiv f(s). \tag{1.4}$$

This is the special form of (1.2) called the *Laplace transform* of $F(t)$ and is also denoted by $£\{F(t)\}$. Thus, we have

$$£\{F(t)\} = \int_{t=0}^{\infty} e^{-st}.F(t)\, dt \equiv f(s). \tag{1.5}$$

§ 2. Fourier transform

The Fourier transform of the function f is traditionally denoted by adding a circumflex: $\hat{f}$. There are several common conventions for defining the Fourier transform of an integrable function $f: \mathbb{R} \rightarrow \mathbb{C}$.

Here we use the following definition:

$$\hat{f}(t) = \int_{-\infty}^{\infty} f(x) \cdot e^{-2i\pi xt} \cdot dt$$

for any real number t. When the independent variable x represents *time*, the transform variable t represents frequency (e.g. if time is measured in seconds, then the frequency is in *hertz*). Under suitable conditions, f is determined by $\hat{f}$ via the **inverse transform**:

$$f(x) = \int_{-\infty}^{\infty} \hat{f}(t) \cdot e^{2i\pi xt} \cdot dt$$

for any real number x.

The reason for the negative sign convention in the definition of $\hat{f}(t)$ is that the integral produces the amplitude and phase of the function $f(x) \cdot e^{-2i\pi xt}$ at frequency zero (0), which is identical to the amplitude and phase of the function $f(x)$ at frequency t, which is what $\hat{f}(t)$ is supposed to represent.

Note 2.1. There is a close connection between the definition of Fourier series and the Fourier transform for functions f that are zero outside an interval. For such a function, we can calculate its Fourier series on any interval that includes the points where f is not identically zero. The Fourier transform is also defined for such a function. As we increase the length of the interval on which we calculate the Fourier series, then the Fourier series coefficients begin to look like the Fourier transform and the sum of the Fourier series of f begins to look like the inverse Fourier transform.

§ 3. Properties of the Fourier transform

Here we consider $f(x)$, $g(x)$ and $h(x)$ as *integrable functions*: Lebesgue–measurable on the real line satisfying:

$$\int_{-\infty}^{\infty} |f(x)| \, dx < \infty.$$

We denote the Fourier transforms of these functions by $\hat{f}(t)$, $\hat{g}(t)$ and $\hat{h}(t)$ respectively.

The Fourier transform has the following basic properties:

(i) Linearity: For any complex numbers a and b, if

$$h(x) = af(x) + bg(x), \qquad \text{then } \hat{h}(t) = a \cdot \hat{f}(t) + b \cdot \hat{g}(t).$$

(ii) Translation / time shifting: For any real number x_0, if

$$h(x) = f(x - x_0), \text{ then} \qquad \hat{h}(t) = \exp(-2\pi i x_0 t). \hat{f}(t).$$

(iii) Modulation / frequency shifting: For any real number t_0, if

$$h(x) = \exp(2\pi i x t_0). f(x), \text{ then} \qquad \hat{h}(t) = \hat{f}(t - t_0).$$

(iv) Time scaling: For a non–zero real number a, if

$$h(x) = f(ax), \qquad \text{then} \qquad \hat{h}(t) = (1/|a|)\hat{f}(t/a).$$

The case $a = -1$ leads to the *time-reversal* property, which states: if

$$h(x) = f(-x), \qquad \text{then} \qquad \hat{h}(t) = \hat{f}(-t).$$

§ 4. Applications of Fourier transform to one-dimensional wave equation

Consider a perfectly flexible elastic string of (natural) length l stretched between two points O ($x = 0$) and A ($x = l$) with uniform tension T. Let the string be displaced slightly from its initial position of rest and released while the end points remain fixed. The string will then vibrate and the vibrations are governed by the one dimensional wave equation written as

$$c^2(\partial^2 u / \partial x^2) = \partial^2 u / \partial t^2. \tag{4.1}$$

Assuming the initial conditions

$$x = 0, \qquad u(0, t) = 0, \tag{4.2a}$$

$$x = l, \qquad u(l, t) = 0, \tag{4.2b}$$

for all values of t and taking the initial deflection and initial velocity as the following

$$u(x, 0) = f(x) \quad (4.3); \qquad (\partial u / \partial t)_{t=0} = g(x), \quad (4.4)$$

we need to find a solution of Eq. (4.1). This is carried out in the following three steps.

Step 1. Applying separation of variables method, let

$$u = X(x). T(t) \tag{4.5}$$

be a trial solution of Eq. (4.1) so that there hold the relations

$$u_x \equiv \partial u / \partial x = X'(x). T(t), \quad u_t \equiv \partial u / \partial t = X(x). T'(t)$$

$$u_{xx} \equiv \partial^2 u / \partial x^2 = X''(x). T(t), \quad u_{tt} \equiv \partial^2 u / \partial t^2 = X(x). T''(t)$$

$$\left. \right\} \tag{4.6}$$

reducing Eq. (4.1) to the form

$$c^2 X''. T = X. T'' \qquad \Rightarrow \qquad X'' / X = T'' / c^2 T. \tag{4.7}$$

The expressions on the two sides of Eq. (4.7) are functions of independent) variables x and t respectively so either of them must be constant, say k. Thus, the equations (4.7) lead to

$$X'' - k. X = 0 \quad (4.8); \quad \text{and} \qquad T'' - c^2. k. T = 0. \tag{4.9}$$

Step 2. We now seek the solutions of these differential equations so that Eq. (4.5) satisfies the initial conditions given by Eq. (4.2) for every value of t :

$$u(0, t) = X(0). T(t) = 0, \tag{4.10a}$$

$$u(l, t) = X(l). T(t) = 0. \tag{4.10b}$$

We discuss both vanishing and non-vanishing choices of the function T: $T = 0$ gives a trivial solution of Eq. (4.1): $u \equiv X. T = 0$, which is of no interest to us. Also, when $T \neq 0$, from Eq. (4.10), there result

$$X(0) = 0, \text{ as well as } X(l) = 0. \tag{4.11}$$

In the following, we will discuss various possibilities for the constant k.

(i) If $k = 0$, the differential equation (4.8) has a general solution $X = a.x + b$; which, for the initial conditions given by Eq. (4.11), determines

$0 = a.0 + b \Rightarrow b = 0$; and $0 = X(l) = a.l + 0 \Rightarrow a = 0$, as $l \neq 0$.

Hence, the solution of Eq. (4.8) becomes $X = 0 \Rightarrow u \equiv X$. $T = 0$ (as before for $T = 0$).

(ii) If $k > 0$, say ρ^2, then Eq. (4.8) assumes the form $X'' - \rho^2 X = 0$ having a solution

$$X = A e^{\rho x} + B e^{-\rho x}. \qquad (4.12)$$

Applying the initial conditions given by Eq. (4.11), we evaluate the constants A and B:

$$X(0) = 0 = A + B \qquad \text{and} \qquad X(l) = 0 = A e^{\rho l} + B e^{-\rho l}.$$

Eliminating B from these we get

$$A.(e^{\rho l} - e^{-\rho l}) = 0 \qquad \Rightarrow \qquad A = 0, \text{ for } e^{\rho l} - e^{-\rho l} \neq 0.$$

Hence, $B = -A = 0$. Thus, Eq. (4.12) again yields $X = 0 \Rightarrow u = 0$ as before.

(iii) When $k < 0$, say $-\rho^2$, then Eq. (4.8) reduces to $X'' + \rho^2 X = 0$ having a general solution

$$X = A \cos \rho x + B \sin \rho x. \qquad (4.13)$$

Applying the initial conditions given by Eq. (4.11), we derive $X(0) = 0 = A$, and $X(l) = B \sin \rho l = 0$. The latter relation gives two alternatives: either $B = 0$, or

$$\sin \rho l = 0. \qquad (4.14)$$

The choice $B = 0$ (together with $A = 0$ as seen above) again reduces Eq. (4.13) to $X = 0$ implying $u = 0$, which is trivial. But, the latter alternative Eq. (4.14) yields $\rho l = 0$, or $n\pi$ (n being an integer). Since $l \neq 0$ and ρ (the mass per unit length of the string) too cannot be zero leaving the possibility

$$\rho = n\pi / l. \qquad (4.15)$$

Setting $B = 1$ (for simplicity), we obtain infinitely large number of solutions of Eq. (4.8):

$$X \equiv X_n(x) = \sin(n\pi x / l), \ n = 1, 2, 3, \ldots \qquad (4.16)$$

For Eq. (4.15), $k \equiv -\rho^2 = -(n\pi / l)^2$, the equation (4.9) takes the form

$$T'' + \lambda_n^2 . T = 0, \tag{4.17}$$

where

$$\lambda_n \equiv c\rho = c\, n\pi / l. \tag{4.18}$$

A general solution of Eq. (4.17) is

$$T_n(t) = A_n . \cos(\lambda_n t) + B_n . \sin(\lambda_n t). \tag{4.19}$$

Thus, Eqs. (4.16) and (4.19) determine solutions of Eq. (4.1):

$$u_n(x, t) \equiv X_n(x) . T_n(t) = \{A_n.\cos(\lambda_n t) + B_n.\sin(\lambda_n t)\} \sin(n\pi x / l). \tag{4.20}$$

Definition 4.1. Above solutions of the wave equation (4.1) are called the *eigen functions* and the values λ_n given by Eq. (4.18) as the *eigen values* of the vibrating string.

Step 3. From Eq. (4.20), we derive

$$u_n(x, 0) = A_n . \sin(n\pi x / l) = f(x), \tag{4.21}$$

for Eq. (4.3); and

$$\partial u_n / \partial t = \{-A_n . \sin(\lambda_n t) + B_n . \cos(\lambda_n t)\} . \lambda_n . \sin(n\pi x / l)$$

$$\Rightarrow$$

$$(\partial u_n / \partial t)_{t=0} = B_n \lambda_n \sin(n\pi x / l) = g(x), \tag{4.22}$$

for Eq. (4.4). Comparing Eqs. (4.21) and (4.22), we note that the functions $f(x)$ and $g(x)$ are not independent and are connected by a linear relation

$$(B_n \lambda_n / A_n) f(x) = g(x).$$

As Eq. (4.1) is linear and homogeneous we cannot accept Eq. (4.20) as a solution. Instead, we consider the infinite series

$$u(x, t) \equiv \sum_{n=1}^{\infty} u_n(x, t)$$

$$= \sum_{n=1}^{\infty} \{A_n . \cos(\lambda_n t) + B_n . \sin(\lambda_n t)\} \sin(n\pi x / l). \tag{4.23}$$

At $t = 0$, we, therefore, have

$$u(x,0) \equiv f(x) = \sum_{n=1}^{\infty} \{A_n. \sin(n\pi x / l). \tag{4.24}$$

This gives an expansion of $f(x)$ in a *sine* series. Hence, in view of Eq. (5.3.16), we get

$$A_n = (2/l) \int_{x=0}^{l} f(x) \sin(n\pi x / l), dx, n = 1, 2, 3, \ldots \tag{4.25}$$

Also, in view of Eq. (4.22), we have

$$(\partial u / \partial t)_{t=0} \equiv \sum_{n=1}^{\infty} (\partial u_n/\partial t)_{t=0} = g(x) = \sum_{n=1}^{\infty} B_n \lambda_n \sin(n\pi x/l). \tag{4.26}$$

This, in analogy with Eq. (4.25), determines

$$B_n \lambda_n = (2/l) \int_{x=0}^{l} g(x). \sin(n\pi x/l). dx,$$

or, for Eq. (4.18),

$$B_n = (2/cn\pi) \int_{x=0}^{l} g(x) \sin(n\pi x / l). dx, \; n = 1, 2, 3, \ldots \tag{4.27}$$

Thus, Eq. (4.23) gives a solution of the wave equation (4.1) where the constants A_n, B_n are determined by Eqs. (4.25) and (4.27).

4.1. Particular case: When the string is released from rest $g(x) = 0$ for every x in the interval $0 \leq x \leq l$, it follows from Eq. (4.27) that $B_n = 0$ reducing the solution vide Eq. (4.23) to

$$u(x, t) = \sum_{n=1}^{\infty} A_n. \cos(\lambda_n t). \sin(n\pi x / l). \tag{4.28}$$

Example 4.1. Let a vibrating string of length 30 cms. satisfies the wave equation

$$4(\partial^2 u / \partial x^2) = \partial^2 u / \partial t^2, \quad 0 < x < 30, \; 0 < t. \tag{4.29}$$

Let the ends of the string be fixed and it is set in motion with zero initial velocity from the initial position:

$$u(x, 0) = f(x) = \begin{cases} x/10, & 0 \leq x \leq 10, \\ (30-x)/20, & 10 \leq x \leq 30. \end{cases} \tag{4.30}$$

Find the displacement $u(x, t)$ of the string.

Solution. A solution of the wave equation is given by equation (4.28). For $c = 2$, $l = 30$ determining $\lambda_n = cn\pi/l = 2n\pi/30 = n\pi/15$, the equation (4.28) becomes

$$u(x, t) = \sum_{n=1}^{\infty} A_n \cos(n\pi t /15) . \sin(n\pi x /30). \qquad (4.31)$$

Also, Eq. (4.25) determines A_n:

$$A_n = (2/30). \int_{x=0}^{30} f(x). \sin(n\pi x / 30). dx$$

$$= (1/15) \{ \int_{x=0}^{10} (x/10). \sin(n\pi x/30). dx$$

$$+ \int_{x=10}^{30} \{(30-x)/20\}. \sin(n\pi x/30). dx \}$$

$$= (1/150) \int_{x=0}^{10} x. \sin(n\pi x/30). dx + (1/10) \int_{x=10}^{30} \sin(n\pi x/30).dx$$

$$- (1/300) \int_{x=10}^{30} x. \sin(n\pi x/30).dx$$

$$= (1/5n\pi) \{ \left[-x. \cos(n\pi x/30)\right]_0^{10} + \int_{x=0}^{10} \cos(n\pi x / 30). dx \}$$

$$+ (3/n\pi) \left[-\cos(n\pi x/30)\right]_{10}^{30}$$

$$+ (1/10n\pi) \{ \left[x. \cos(n\pi x/30)\right]_{10}^{30} - \int_{x=10}^{30} \cos(n\pi x / 30). dx \}$$

$$= (-2/n\pi) \cos(n\pi/3) + (6/n^2\pi^2) \left[\sin(n\pi x/30)\right]_0^{10}$$

$$+ (3/n\pi)\{ -\cos n\pi + \cos(n\pi/3)\} + (1/n\pi)\{3 \cos n\pi - \cos(n\pi/3)\}$$

$$- (3/n^2\pi^2) \left[\sin(n\pi x/30)\right]_{10}^{30}$$

$$= (6/n^2\pi^2) \sin(n\pi/3) - (3/n^2\pi^2) \{\sin n\pi - \sin(n\pi/3)\}$$

$$= (9/n^2\pi^2) \sin(n\pi/3). //$$

CHAPTER 8

FOURIER SERIES

§ 1. Fourier series

Let $f(x)$ be some function of x defined over some interval, say $[-\pi, \pi]$ of the real line and it satisfies certain conditions. We know, by Taylor's theorem of differential calculus, the function can be expanded in infinite series of terms in polynomial form about a given point x_0 in the interval:

$$f(x) = \sum_{n=0}^{\infty} a_n (x - x_0)^n. \tag{1.1}$$

Expansion of $f(x)$ in an infinite series whose terms need not be polynomials is also possible. One of the most useful such expansion is the following trigonometric series:

$$f(x) = a_0 + (a_1 \cos x + a_2 \cos 2x + \ldots + a_n \cos nx + \ldots)$$

$$+ (b_1 \sin x + b_2 \sin 2x + \ldots + b_n \sin nx + \ldots)$$

$$= a_0 + \sum_{n=1}^{\infty} (a_n \cos nx + b_n \sin nx), \tag{1.2}$$

so that each term in the series is term-wise integrable. Above series is called a *Fourier series* for the function $f(x)$ over the given interval $[-\pi, \pi]$. The coefficients $a_0, a_1, a_2, \ldots, a_n; b_1, b_2, \ldots, b_n$ are called the *Fourier coefficients*.

§ 2. Evaluation of Fourier coefficients

2.1. Integrating the series in Eq. (1.2) over the interval $[-\pi, \pi]$, and applying the properties of definite integrals given by Eq. (1.1.7):

$$\int_{-\pi}^{\pi} \cos nx \, dx = 2 \int_{0}^{\pi} \cos nx \, dx = (2/n) \left[\sin nx \right]_0^\pi = 0 \tag{2.1}$$

$$\int_{-\pi}^{\pi} \sin nx \, dx = 0, \ (\sin x \text{ being an odd function of } x) \tag{2.2}$$

we evaluate the Fourier coefficients:

$$\int_{-\pi}^{\pi} f(x)\, dx = a_o \int_{-\pi}^{\pi} 1\, dx = 2a_o \int_{0}^{\pi} 1\, dx = 2\pi a_o$$

$$\Rightarrow \qquad a_o = (1/2\pi)\int_{-\pi}^{\pi} f(x)\, dx. \qquad (2.3)$$

2.2. Multiplying the series in Eq. (1.2) by cos *mx* and integrating the result so obtained over the interval $[-\pi, \pi]$, we get

$$\int_{-\pi}^{\pi} f(x).\cos mx.\, dx = a_o.\int_{-\pi}^{\pi} \cos mx.\, dx$$

$$+ \sum_{n=1}^{\infty} a_n.\int_{-\pi}^{\pi} \cos mx.\cos nx.\, dx + \sum_{n=1}^{\infty} b_n \int_{-\pi}^{\pi} \cos mx.\sin nx\, dx. \qquad (2.4)$$

The integrands cos *mx*. cos *nx* (respectively cos *mx*. sin *nx*) being even (resp. odd) functions, for Eqs. (2.1), (2.2) and the properties of definite integrals given by Eq. (1.1.7), Eq. (2.4) simplifies to

$$\int_{-\pi}^{\pi} f(x).\cos mx\, dx = \sum_{n=1}^{\infty} a_n \int_{0}^{\pi} \{\cos (m-n)\, x + \cos (m+n)\, x\, dx \qquad (2.5)$$

$$= \sum_{n=1}^{\infty} a_n \big[\{\sin (m-n)x\}/(m-n) + \{\sin (m+n)x\}/(m+n)\big]_{0}^{\pi} = 0,$$

$$\text{when } n \neq m \quad (2.6)$$

On the other hand, when $n = m$, Eq. (2.5) determines

$$\int_{-\pi}^{\pi} f(x) \cos mx\, dx = a_m \int_{0}^{\pi} \{1 + \cos 2mx)\, dx$$

$$= a_m \big[x + (\sin 2mx)/2m\big]_{0}^{\pi} = \pi a_m$$

$$\Rightarrow \qquad a_m = (1/\pi) \int_{-\pi}^{\pi} f(x).\cos mx\, dx. \qquad (2.7)$$

2.3. Multiplying the series in Eq. (1.2) by sin *mx* and integrating the result so obtained over the interval $(-\pi, \pi)$, we also find

$$\int_{-\pi}^{\pi} f(x) \sin mx\, dx = a_o \int_{-\pi}^{\pi} \sin mx + \sum_{n=1}^{\infty} a_n \int_{-\pi}^{\pi} \sin mx.\cos nx\, dx$$

$$+ b_n \int_{-\pi}^{\pi} \sin mx.\sin nx\, dx.$$

Again, for Eq. (2.2) and the properties of definite integrals given by Eq. (1.1.7), above equation simplifies to

$$\int_{-\pi}^{\pi} f(x).\sin mx \, dx = \sum_{n=1}^{\infty} b_n \int_0^{\pi} \{\cos(m-n)x - \cos(m+n)x \, dx \quad (2.8)$$

$$= \sum_{n=1}^{\infty} b_n \left[\{\sin(m-n)x\}/(m-n) - \{\sin(m+n)x\}/(m+n)\right]_0^{\pi} = 0,$$

when $n \neq m$. On the other hand, when $n = m$, Eq. (2.8) determines

$$\int_{-\pi}^{\pi} f(x) \sin mx \, dx = b_m \int_0^{\pi} (1 - \cos 2mx) \, dx$$

$$= b_m \left[x - (\sin 2mx)/2m\right]_0^{\pi} = \pi b_m$$

$$\Rightarrow \qquad b_m = (1/\pi) \int_{-\pi}^{\pi} f(x).\sin mx \, dx, \qquad (2.9)$$

Putting from Eqs. (2.3), (2.7) and (2.9), the Fourier series in Eq. (1.2) for the function $f(x)$ assumes the form:

$$f(x) = (1/2\pi)\int_{-\pi}^{\pi} f(x) \, dx + (1/\pi) \sum_{n=1}^{\infty} [\{\int_{-\pi}^{\pi} f(x) \cos nx. \, dx \} \cos nx$$

$$+ \{\int_{-\pi}^{\pi} f(x).\sin nx. \, dx\}. \sin nx \,] \qquad (2.10)$$

$$= (1/2\pi)\int_{-\pi}^{\pi} f(t) \, dt + (1/\pi) \sum_{n=1}^{\infty} \int_{-\pi}^{\pi} f(t) \{ \cos nt. \cos nx$$

$$+ \sin nt. \sin nx \} \, dt$$

$$= (1/2\pi)\int_{-\pi}^{\pi} f(t) \, dt + (1/\pi) \sum_{n=1}^{\infty} \int_{-\pi}^{\pi} f(t).\cos \{n(t-x)\} \, dt. \quad (2.11)$$

2.4. Particular cases

(i) When $f(x)$ is an even function of x, i.e. $f(-x) = f(x)$, the Fourier coefficients reduce to

$$a_o = (1/\pi)\int_0^{\pi} f(x) \, dx, \quad a_n = (2/\pi)\int_0^{\pi} f(x).\cos nx. \, dx, \quad b_n = 0, \quad (2.12)$$

for the odd character of the integrand in Eq. (2.9) causing vanishing of the integral. Consequently, the Fourier series given by Eq. (2.10), for such a function, reduces to

$$f(x) = a_o + \sum_{n=1}^{\infty} a_n \cos nx = (1/\pi) [\int_0^{\pi} f(x)\, dx$$

$$+ 2 \sum_{n=1}^{\infty} \{\int_0^{\pi} f(x). \cos nx\, dx\} \cos nx] \qquad (2.10a)$$

(ii) On the other hand, if $f(x)$ is an odd function of x, i.e. $f(-x) = -f(x)$, the Fourier coefficients reduce to

$$a_o = 0, \quad a_n = 0, \quad b_n = (2/\pi) \int_0^{\pi} f(x). \sin nx\, dx \qquad (2.13)$$

again for the odd character of the integrand in Eq. (2.7) causing vanishing of the integral. Consequently, the Fourier series given by Eq. (2.10), for an odd function, reduces to

$$f(x) = \sum_{n=1}^{\infty} b_n \sin nx = (2/\pi) \sum_{n=1}^{\infty} \{\int_0^{\pi} f(x). \sin nx\, dx\}. \sin nx]. \quad (2.10b)$$

Thus, we conclude that:

(i) The Fourier series for an *even* function consists of a constant term a_o and the *cosine* series;

(ii) The Fourier series for an *odd* function consists of only a *sine* series.

Example 2.1. Find the Fourier series for the function $f(x) = x$ defined in the interval $[-\pi, \pi]$.

Solution. The function being odd, the Fourier coefficients can be evaluated by (2.13): $a_o = 0$, $a_n = 0$, and

$$b_n = (2/\pi) \int_0^{\pi} x. \sin nx. \, dx = (2/n\pi)\{ [-x.\cos nx]_0^{\pi} + \int_0^{\pi} \cos nx. \, dx \}$$

$$= (2/n\pi) \{ -\pi. \cos n\pi + (1/n).[\sin nx]_0^{\pi} \}$$

$$= (2/n\pi)\{ -\pi. \cos n\pi + (1/n) \sin n\pi \} = -(2/n). \cos n\pi = (2/n) (-1)^{n+1}.$$

Hence, the Fourier series in Eq. (1.2) reduces to

$$f(x) \equiv x = \sum_{n=1}^{\infty} (2/n)(-1)^{n+1} . \sin nx$$

$$= 2\{(\sin x)/1 - (\sin 2x)/2 + (\sin 3x)/3 - (\sin 4x)/4 + \dots \infty\}. \text{ // } (2.14)$$

Example 2.2. Find a series of *sines* and *cosines* of multiples of x which shall represent the function $f(x) = x + x^2$ in the interval $[-\pi, \pi]$. Hence, show that

$$\pi^2/6 = 1 + 1/2^2 + 1/3^2 + \dots \infty. \tag{2.15}$$

Solution. (i) Eqs. (2.3), (2.7) and (2.9) determine the Fourier coefficients:

$$a_0 = (1/2\pi) \int_{-\pi}^{\pi} (x + x^2) \, dx = (1/2\pi) \{ \int_{-\pi}^{\pi} x. \, dx + \int_{-\pi}^{\pi} x^2. \, dx \}$$

$$= (1/\pi) \{0 + \int_{0}^{\pi} x^2. \, dx\} = (1/\pi) \left[x^3/3\right]_{0}^{\pi} = \pi^2/3,$$

$$a_n = (1/\pi) \int_{-\pi}^{\pi} (x + x^2). \cos nx. dx$$

$$= (1/\pi) \{ \int_{-\pi}^{\pi} x. \cos nx. \, dx + \int_{-\pi}^{\pi} x^2. \cos nx. \, dx\}$$

$$= (2/\pi) \{ 0 + \int_{0}^{\pi} x^2. \cos nx. dx \}$$

$$= (2/\pi) \{ \left[(x^2/n). \sin nx\right]_{0}^{\pi} - (2/n) \int_{0}^{\pi} x. \sin nx. \, dx \}$$

$$= (2\pi/n). \sin n\pi + (4/n\pi) \{ \left[(x/n). \cos nx\right]_{0}^{\pi} - (1/n) \int_{0}^{\pi} \cos nx. \, dx\}$$

$$= (4/n^2). \cos n\pi - (4/n^3 \pi) \left[\sin nx\right]_{0}^{\pi} = (4/n^2). \cos n\pi = (4/n^2).(-1)^n,$$

and

$$b_n = (1/\pi) \int_{-\pi}^{\pi} (x + x^2). \sin nx. dx$$

$$= (1/\pi) \{ \int_{-\pi}^{\pi} x. \sin nx. \, dx + \int_{-\pi}^{\pi} x^2. \sin nx. \, dx\}$$

$$= (2/\pi) \ \{ \int_0^\pi x. \sin nx. \ dx + 0 \} = (2/n\pi)\{ \left[-x.\cos nx \right]_0^\pi + \int_0^\pi \cos nx \ dx \}$$

$$= -(2/n). \cos n\pi + (2/n^2 \pi) \left[\sin nx \right]_0^\pi = -(2/n). \cos n\pi = (2/n).(-1)^{n+1}.$$

Accordingly, the Fourier series in Eq. (1.2) reduces to

$$x + x^2 = \pi^2/3 + 4 \ \{-(\cos x)/1^2 + (\cos 2x)/2^2 - (\cos 3x)/3^2 + \dots \infty\}$$

$$+ 2 \ \{(\sin x)/1 - (\sin 2x)/2 + (\sin 3x)/3 - \dots \infty\}. \tag{2.16}$$

(ii) Similarly, the Fourier series for the function x^2 in above interval
is
$$x^2 = \pi^2/3 + 4\{-(\cos x)/1^2 + (\cos 2x)/2^2 - (\cos 3x)/3^2 + \dots \infty\}. \tag{2.17}$$

Putting $x = \pi$ in above result, we deduce

$$\pi^2 - \pi^2/3 = 2\pi^2/3 = 4\{1 + 1/2^2 + 1/3^2 + \dots \infty\},$$

which yields (2.15). //

Example 2.3. Obtain a Fourier series expansion of the function $f(x)$
$= x. \sin x$ in the interval $[-\pi, \pi]$. Hence, deduce the result

$$\pi/4 = 1/2 + (1/1.3 - 1/3.5 + 1/5.7 - \dots \infty). \tag{2.18}$$

Solution. (i) The function being even, the Fourier coefficients are
evaluated by Eq. (2.12):

$$a_0 = (1/\pi) \ \{ \int_0^\pi x. \sin x. \ dx \} = (1/\pi) \ \{ \left[-x.\cos x \right]_0^\pi + \int_0^\pi \cos x. dx \}$$

$$= -\cos \pi + (1/\pi) \left[\sin x \right]_0^\pi = 1,$$

$$a_n = (2/\pi). \int_0^\pi x. \sin x. \cos nx. \ dx$$

$$= (1/\pi) \int_0^\pi x.\{\sin (n+1) x - \sin (n-1) x\}. \ dx$$

$$= (1/\pi) \ \{ \int_0^\pi x. \sin (n+1) x. \ dx - \int_0^\pi x. \sin (n-1) x\}. \ dx \tag{2.19}$$

$$= \{ \left[-x.\cos\ (n+1)\ x \right]_0^\pi + \int_0^\pi \cos\ (n+1)\ x.\ dx \} \ / \ (n+1)\ \pi$$

$$+ \{ \left[x.\cos\ (n-1)x \right]_0^\pi - \int_0^\pi \cos\ (n-1)\ x.\ dx \} \ / \ (n-1)\ \pi, \qquad \text{when } n \neq 1,$$

$$= \left[- \{\cos\ (n+1)\ \pi\}/\ (n+1) + \{\sin\ (n+1)\ \pi\}/\ (n+1)^2\ \pi \right]$$

$$+ \left[\{\cos\ (n-1)\ \pi\}/(n-1) - \{\sin\ (n-1)\ \pi\}/\ (n-1)^2\ \pi \right]$$

$$= (\cos\ n\pi)\ \{1/\ (n+1) - 1/\ (n-1)\} \ = \ -2\ (\cos\ n\pi)\ /\ (n-1)\ (n+1)$$

$$= \ 2\ (-1)^{\,n+1}/\ (n-1)\ (n+1),$$

and $b_n = 0$. However, when $n = 1$, Eq. (2.19) determines

$$a_1 \ = \ (1/\pi).\ \int_0^\pi x.\ \sin\ 2x.\ dx$$

$$= \ (1/2\pi)\ \{ \left[-x.\cos\ 2x \right]_0^\pi + \int_0^\pi \cos\ 2x.\ dx \}$$

$$= \ -1/2\ +(1/4\pi) \left[\sin\ 2x \right]_0^\pi \ = \ -1/2.$$

Accordingly, the Fourier series in Eq. (1.2) reduces to

$$x.\ \sin\ x \ = \ 1 - (\cos\ x)/2 + 2 \sum_{n=2}^{\infty} \ \{(-1)^{\,n+1} \cos\ nx\} \ / \ (n-1)(n+1)$$

$$= 1 - (\cos\ x)/2 + 2\ \{(-\cos\ 2x)/1.3 + (\cos\ 3x)/2.4 - (\cos\ 4x)/3.5$$

$$+ (\cos\ 5x)\ /\ 4.6 - (\cos\ 6x)\ /\ 5.7 + ...\ \infty\}. \qquad (2.20)$$

(ii) Putting $x = \pi/2$ in above result, we deduce

$$\pi/2 \ = \ 1 + 2\ \{1/\ 1.3 + 0 - 1/\ 3.5 + 0 +1/\ 5.7 - 0 + ...\ \infty\},$$

that yields Eq. (2.18). //

Example 2.4. In the interval $[-\pi,\ \pi]$, prove that

$$x.\ (\pi^2 - x^2)\ /\ 12 = (\sin\ x)/1^3 - (\sin\ 2x)/2^3 + (\sin\ 3x)/3^3 - ...\ \infty\}. \qquad (2.21)$$

Solution. The function $x (\pi^2 - x^2) / 12$ being odd the Fourier coefficients are evaluated by Eq. (2.13):

$$a_0 = a_n = 0, \qquad b_n = (2/12\pi) \int_0^\pi x. (\pi^2 - x^2). \sin nx. \, dx.$$

Carrying out the integration by the method of parts taking $\sin nx$ as the second function we obtain

$$b_n = (1/6n\pi). \left\{ \left[-x.(\pi^2 - x^2).\cos nx \right]_0^\pi + \int_0^\pi (\pi^2 - 3x^2). \cos nx \, dx \right\}$$

$$= (1/6n^2 \pi). \left\{ \left[(\pi^2 - 3x^2).\sin nx \right]_0^\pi + 6 \int_0^\pi x. \sin nx \, dx \right\}$$

$$= (1/n^3 \pi). \left\{ \left[-x.\cos nx \right]_0^\pi + \int_0^\pi \cos nx. \, dx \right\}$$

$$= (1/n^3) \left\{ - \cos n\pi + (1/n\pi) \left[\sin nx \right]_0^\pi \right\} = -(1/n^3) \cos n\pi = (-1)^{n+1}/n^3.$$

Accordingly, the Fourier series in Eq. (1.2) for the function reads as

$$x. (\pi^2 - x^2) / 12 = \sum_{n=1}^\infty \left\{ (-1)^{n+1} / n^3 \right\}. \sin nx$$

$$= (\sin x) / 1^3 - (\sin 2x) / 2^3 + (\sin 3x) / 3^3 - \ldots \infty \}. \, //$$

§ 3. Fourier series in any interval

3.1. Interval $[-l, l]$: The equation (2.11) gives a Fourier series expansion for the function $f(x)$ in the interval $(-\pi, \pi)$. Analogously, for some function $F(y)$ also defined in the same interval $[-\pi, \pi]$, Eq. (2.11) reads as

$$F(y) = (1/2\pi) \int_{-\pi}^\pi F(t) \, dt + (1/\pi) \sum_{n=1}^\infty \int_{-\pi}^\pi F(t) \cos \{n(t-y)\} dt. \ (3.1)$$

Applying the change of parameters:

$$t = \pi u / l \quad \Rightarrow \quad dt = (\pi. \, du) / l, \quad \text{and} \quad y = \pi x / l. \ (3.2)$$

Eq. (3.1) assumes the form

$$F(\pi x / l) = (1/2l) \int_{-l}^l F(\pi u / l) \, du$$

$$+ (1/l) \sum_{n=1}^{\infty} \int_{-l}^{l} F\,(\pi u\,/\,l)\cos\,\{n\pi\,(u-x)\,/\,l\}.\,du$$

Writing $F\,(\pi x\,/\,l)$ as $f(x)$, and similarly $F\,(\pi u\,/\,l)$ as $f(u)$ above relation further alters as

$$f(x) = (1/2l)\int_{-l}^{l} f(u)\,du + (1/\,l)\sum_{n=1}^{\infty}\int_{-l}^{l} f(u)\cos\,\{n\pi\,(u-x)\,/\,l\}.du$$

$$= (1/2l)\int_{-l}^{l} f(u)\,du + (1/\,l)\sum_{n=1}^{\infty}\int_{-l}^{l} f(u)\,\{\cos\,(n\pi u/\,l)\cos\,(n\pi x/\,l)$$

$$+\sin\,(n\pi u\,/\,l)\sin\,(n\pi x\,/\,l\,)\}\,du, \tag{3.3}$$

which is the desired Fourier series expansion of $f(x)$ in interval $[-\,l,\,l\,]$ together with the Fourier coefficients:

$$a_0 = (1/2l)\int_{-l}^{l} f(u)\,du,\; a_n = (1/l)\int_{-l}^{l} f(u)\cos\,(n\pi u/l)\,du$$

$$\left.\vphantom{\int}\right\} \tag{3.4}$$

$$b_n = (1/l)\int_{-l}^{l} f(u)\sin\,(n\pi u\,/\,l\,)\,du.$$

3.2. Interval $[0,\,\pi]$: Let a function $f(x)$ be defined on the interval $[0,\,\pi]$. In order to obtain a *cosine* series expansion for it on this interval, we construct a new function, say $g(x)$ on the extended interval $[-\,\pi,\,\pi]$ as follows:

$$g(x) = f(-x)\;(\text{when }-\pi \le x < 0),\text{ and } f(x)\;(\text{when } 0 \le x \le \pi). \tag{3.5}$$

As such, the new function is even on the extended interval and agrees with the given function $f(x)$ on the interval $[0,\,\pi]$. Thus, $g(x)$ can be expanded in a Fourier series as discussed in Sub-section 2.4, Case (i). The Fourier coefficients for the same are

$$a_0 = (1/2\pi)\int_{-\pi}^{\pi} g(x)\,dx = (1/\,\pi)\int_{0}^{\pi} g(x)\,dx = (1/\pi)\int_{0}^{\pi} f(x)\,dx$$

$$a_n = (1/\pi)\int_{-\pi}^{\pi} g(x)\cos\,nx.\,dx = (2/\pi)\int_{0}^{\pi} g(x).\cos\,nx.\,dx$$

$$= (2\,/\,\pi)\int_{0}^{\pi} f(x).\cos\,nx\,dx,$$

and

$$b_n = (1/\pi) \int_{-\pi}^{\pi} g(x). \sin nx \, dx = 0,$$

as the integrand $g(x). \sin nx$ then becomes an odd function of x causing vanishing of the last integral. We note that above values of the Fourier coefficients are in accordance with Eq. (2.12). Consequently, the Fourier series for $g(x)$ on the interval $[-\pi, \pi]$ turns out to be a purely *cosine* series given by Eq. (2.10a) on $[-\pi, \pi]$. For $g(x)$ being same as $f(x)$ on the interval $[0, \pi]$ the same *cosine* series is valid for $f(x)$ too on the interval $[0, \pi]$.

Example 3.1. Find a series of cosines of multiples of x which may represent the function $f(x) = x$ on the interval $[0, \pi]$.

Solution. To have a cosine series only the function must be so defined that it be *even* in the extended interval $[-\pi, \pi]$. Therefore, as seen above the Fourier coefficients are determined as per Eq. (2.12):

$$a_0 = (1/\pi) \int_0^{\pi} f(x) \, dx = (1/\pi) \int_0^{\pi} x \, dx = (1/\pi) \left[x^2/2 \right]_0^{\pi} = \pi/2,$$

$$a_n = (2/\pi) \int_0^{\pi} x \cos nx. \, dx = (2/n\pi)\{\left[x. \sin nx \right]_0^{\pi} - \int_0^{\pi} \sin nx. \, dx\}$$

$$= (2/n^2 \pi) \left[\cos nx \right]_0^{\pi} = (2/n^2 \pi)(\cos n\pi - 1) = (2/n^2 \pi)\{(-1)^n - 1\}$$

$$= 0 \text{ (when } n \text{ is even)}, \quad \text{and} -4/n^2 \pi \quad \text{(when } n \text{ is odd)},$$

and $b_n = 0$. Thus, the Fourier series reducing to cosine series, given by Eq. (2.10a), reads as

$$x = \pi/2 - (4/\pi).\{(\cos x)/1^2 + (\cos 3x)/3^2 + (\cos 5x)/5^2 + \dots \infty\}. \quad (3.6)$$

Particularly, when $x = 0$, above series implies the result

$$\pi^2/8 = 1/1^2 + 1/3^2 + 1/5^2 + \dots \infty. \, //$$

Example 3.2. Obtain a cosine series for the function $f(x) = \sin x$ on the interval $[0, \pi]$.

Solution. In order to have a cosine series expansion the function must be so defined that it be *even* in the extended interval $[-\pi, \pi]$.

Therefore, as seen above the Fourier coefficients are determined as per Eq. (2.12):

$$a_0 = (1/\pi) \int_0^\pi \sin x . \, dx = (1/\pi) \left[-\cos x \right]_0^\pi = (1/\pi)(-\cos \pi + 1) = 2/\pi,$$

$$a_n = (2/\pi) \int_0^\pi \sin x . \cos nx . \, dx$$

$$= (1/\pi) \int_0^\pi \{ \sin (n+1) x - \sin (n-1) x \} \, dx \qquad (3.7)$$

$$= (1/\pi) \left[\{ -\cos (n+1) x \} / (n+1) + \{ \cos (n-1) x \} / (n-1) \right]_0^\pi, \text{ when } n \neq 1$$

$$= (1/\pi) \left[-\{ \cos (n+1) \pi \} / (n+1) + \{ \cos (n-1) \pi \} / (n-1) + \{ 1/(n+1) - 1/(n-1) \} \right]$$

$$= (1/\pi) \left[(\cos n \pi) / (n+1) - (\cos n \pi) / (n-1) + \{ 1/(n+1) - 1/(n-1) \} \right]$$

$$= (1/\pi) \left[\{ (-1)^n + 1 \} \{ 1/(n+1) - 1/(n-1) \} \right] = -2 \{ (-1)^n + 1 \} / (n-1)(n+1) \, \pi$$

$$= 0 \text{ (when } n \text{ is odd), and } -4 / (n-1) (n+1) \, \pi \text{ (when } n \text{ is even),}$$

and $b_n = 0$.

On the other hand, when $n = 1$, Eq. (3.7) determines

$$a_1 = (1/\pi) \int_0^\pi \sin 2x . \, dx = (1/2\pi) \left[-\cos 2x \right]_0^\pi = (-1 + 1) / 2\pi = 0.$$

Accordingly, the *cosine* series given by Eq. (2.10a) assumes the form

$$\sin x = 2/\pi - (2/\pi) \sum_{n=2}^{\infty} [\{ (-1)^n + 1 \} / (n-1)(n+1)] \cos nx, \, n \text{ being even}$$

$$= 2/\pi - (4/\pi) \sum_{m=1}^{\infty} (\cos 2mx) / (2m-1) (2m+1), \text{ where } n = 2m$$

$$= 2/\pi - (4/\pi) \{ (\cos 2x) / 1.3 + (\cos 4x) / 3.5 + (\cos 6x) / 5.7$$

$$+ (\cos 8x) / 7.9 + \dots \infty \}. \; //$$

Example 3.3. Obtain the Fourier series for the function $f(x) = \cos x$ for all values of x in the interval $[0, \pi]$.

Solution. We construct a new function $g(x)$ in the extended interval $[-\pi, \pi]$ such that

$$g(x) = -\cos x \quad (\text{when } -\pi \leq x \leq 0), \quad \cos(x) \quad (\text{when } 0 \leq x \leq \pi).$$

As per definition it is an odd function of x for which the Fourier coefficients are given by Eq. (2.13): $a_0 = a_n = 0$, and

$$b_n = (2/\pi) \int_0^\pi \cos x . \sin nx . dx$$

$$= (1/\pi) \int_0^\pi \{\sin(n+1)x + \sin(n-1)x\} . dx \qquad (3.8)$$

$$= (1/\pi)\left[-\{\cos(n+1)x\}/(n+1) - \{\cos(n-1)x\}/(n-1)\right]_0^\pi, \text{ when } n \neq 1$$

$$= \left[-\{\cos(n+1)\pi\}/(n+1) - \{\cos(n-1)\pi\}/(n-1) + \{1/(n+1) + 1/(n-1)\}\right]/\pi$$

$$= (1/\pi)\left[(\cos n\pi)/(n+1) + (\cos n\pi)/(n-1) + 1/(n+1) + 1/(n-1)\right]$$

$$= (1/\pi)\left[\{(-1)^n + 1\}\{1/(n+1) + 1/(n-1)\}\right]$$

$$= (2n/\pi)\{(-1)^n + 1\}/(n-1)(n+1)$$

$= 0$ (when n is odd), and $4/(n-1)(n+1)\pi$ (when n is even).   (3.9)

On the other hand, when $n = 1$, Eq. (3.8) determines

$$b_1 = (1/\pi) \int_0^\pi \sin 2x . dx = (1/2\pi)\left[-\cos 2x\right]_0^\pi = (-1+1)/2\pi = 0.$$

Accordingly, the cosine series given by Eq. (2.10b) assumes the form

$$\cos x = \sum_{n=2}^\infty b_n . \sin nx = (4/\pi) \sum_{n=2}^\infty (n \sin nx)/(n-1)(n+1), \text{ by Eq. (3.9)}$$

(where only even values of n give non-zero terms)

$$= (4/\pi)\{2(\sin 2x)/1.3 + 4(\sin 4x)/3.5 + 6(\sin 6x)/5.7 + \ldots \infty\}. //$$

Note 3.1. Defining $g(x)$ as even function on the interval $[-\pi, \pi]$ in analogy with Eq. (3.5), we note that all the Fourier coefficients vanish for $\cos x$ except a_1:

$$a_o = (1/\pi) \int_0^\pi \cos x . \, dx = (1/\pi) \left[\sin x \right]_0^\pi = 0,$$

$$a_n = (2/\pi) \int_0^\pi \cos x . \cos nx . \, dx$$

$$= (1/\pi) \int_0^\pi \{\cos (n + 1) \, x + \cos (n - 1) \, x\} \, dx$$

$$= (1/\pi) \left[\{ \sin (n+1) \, x \} / (n+1) + \{\sin (n-1) \, x\} / (n-1) \right]_0^\pi = 0, \text{ when } n \neq 1;$$

$$a_1 = (1/\pi) \int_0^\pi (\cos 2x + 1) \, dx = (1/\pi) \left[(\sin 2x) / 2 + x \right]_0^\pi = 1$$

and $b_n = 0$. Thus, a *cosine* series for $\cos x$ reduces to just $\cos x$.

3.3. A sine series in the interval [0, π]: To obtain a *sine* series expansion for a function $f(x)$ defined on the interval $[0, \pi]$, we must construct a new function, say $g(x)$ on the extended interval $[-\pi, \pi]$ which should be odd:

$$g(x) = f(x) \text{ (when } 0 \leq x \leq \pi) \text{ and } -f(x) \text{ (when } -\pi \leq x < 0), \quad (3.10)$$

and agree with the given function $f(x)$ on the interval $[0, \pi]$. The Fourier coefficients for the same are given by Eq. (2.13). Accordingly, its Fourier series expansion turns out to be purely a *sine* series as in Eq. (2.10b). Since $g(x)$, as per definition, agrees with $f(x)$ on the interval $[0, \pi]$ its Fourier series in the interval $[0, \pi]$ becomes the same as for $f(x)$.

Example 3.4. Expand the function $f(x) = \sin x$ in a sine series on the interval $[0, \pi]$.

Solution. We construct a new function $g(x)$ on the extended interval $[-\pi, \pi]$ as in Eq. (3.10). The function being odd in $[-\pi, \pi]$ its Fourier series becomes purely a *sine* series given by Eq. (2.10b):

$$g(x) = \sum_{n=1}^\infty b_n . \sin nx \qquad (3.11)$$

and the coefficients b_n are given by Eq. (2.13):

$$b_n = (2/\pi) \int_0^\pi g(x) . \sin nx . dx.$$

For $g(x) = f(x)$ on the interval $[0, \pi]$, above value becomes

$$b_n = (2/\pi) \int_0^\pi f(x) . \sin nx . dx = (2/\pi) \int_0^\pi \sin x . \sin nx . dx$$

$$= (1/\pi) \int_0^\pi \{\cos(n-1)x - \cos(n+1)x\} . dx \qquad (3.12)$$

$$= (1/\pi) \Big[\{\sin(n-1)x\}/(n-1) - \{\sin(n+1)x\}/(n+1) \Big]_0^\pi = 0, \text{ when } n \neq 1.$$

On the other hand, when $n = 1$, (3.12) determines

$$b_1 = (1/\pi) \int_0^\pi (1 - \cos 2x) . dx = (1/\pi) \Big[x - (\sin 2x)/2 \Big]_0^\pi = 1.$$

Thus, the RHS of Eq. (3.11) reduces to just one term, $\sin x$, which is the value of the function itself. //

3.4. The interval $[0, l\,]$: As seen in the Sub-section 3.2, a function can be expanded in the interval $[0, \pi]$ in a *cosine* series of the form in Eq. (2.10a). Re-writing it for a function $F(y)$:

$$F(y) = (1/\pi) \Big[\int_0^\pi F(t) \, dt + 2 \sum_{n=1}^\infty \{ \int_0^\pi F(t) \cos nt . dt \} \cos ny \Big]. \quad (3.13)$$

Applying the change of parameters as in Eq. (3.2), above result assumes the form:

$$F(\pi x / l) = (1/l) [\int_0^l F(\pi u / l) \, du$$

$$+ 2 \sum_{n=1}^\infty \{ \int_0^l F(\pi u / l) . \cos(n\pi u / l) \, du \} . \cos(n\pi x / l)]$$

Writing $F(\pi x/l) \equiv f(x)$, and $F(\pi u/l) \equiv f(u)$, above relation further alters as

$$f(x) = (1/l) [\int_0^l f(u) \, du$$

$$+ 2 \sum_{n=1}^\infty \{ \int_0^l f(u) \cos(n\pi u / l) \, du \} \cos(n\pi x / l)], \quad (3.14)$$

which is a *cosine* series.

The *sine* series expansion of a function $f(x)$ in the interval $[0, \pi]$ is discussed in the Sub-section 3.3 and it is given by Eq. (2.10b). Rewriting it for a function $F(y)$:

$$F(y) = (2/\pi) \sum_{n=1}^{\infty} \{\int_0^{\pi} F(t) \sin nt. \, dt\}. \sin ny, \qquad (3.15)$$

and applying the change of parameters given by Eq. (3.2), above equation reduces to

$$F(\pi x / l) = (2/l) \sum_{n=1}^{\infty} \{\int_0^l F(\pi u/l) \sin (n\pi u/l) \, du\} \sin (n\pi x/l)$$

or, $f(x) = (2/l) \sum_{n=1}^{\infty} \{\int_0^l f(u) \sin (n\pi u / l) \, du\} \sin (n\pi x / l). \; // \quad (3.16)$

Example 3.5. For all values of x in the interval $[-\pi/2, \pi/2]$, prove

$$x = (4/\pi) \{ (\sin x)/1^2 - (\sin 3x)/3^2 + (\sin 5x)/5^2 - \ldots \infty \}. \quad (3.17)$$

Solution. Setting

$$x + \pi/2 = z \qquad (3.18)$$

so that when x varies from $-\pi/2$ to $\pi/2$ the new variable z varies from 0 to π. The *cosine* series expansion of x in the interval $[0, \pi]$ is obtained vide Eq. (3.6). Analogously, we can write a *cosine* series for z in the interval $[0, \pi]$:

$$z = \pi/2 - (4/\pi) \{ (\cos z)/1^2 + (\cos 3z)/3^2 + (\cos 5z)/5^2 + \ldots \infty \},$$

which, for Eq. (3.18), takes the form as given by Eq. (3.17). //

3.5. The interval $[0, 2\pi]$: Putting $t = u - \pi$ so that $dt = du$ and $y = x - \pi$ in Eq. (3.1) we deduce

$$F(x - \pi) = (1/2\pi) \int_{u=0}^{2\pi} F(u - \pi) \, du$$

$$+ (1/\pi) \sum_{n=1}^{\infty} \int_{u=0}^{2\pi} F(u - \pi) \cos \{n(u - x)\} du.$$

Writing $F(x - \pi)$ as $f(x)$, and $F(u - \pi)$ as $f(u)$ above relation further alters as

$$f(x) = (1/2\pi) \int_{u=0}^{2\pi} f(u)\, du$$

$$+ (1/\pi) \sum_{n=1}^{\infty} \int_{u=0}^{2\pi} f(u) \cos \{n\,(u-x)\}\, du \qquad (3.19a)$$

$$= (1/2\pi) \int_{u=0}^{2\pi} f(u)\, du$$

$$+ (1/\pi) \sum_{n=1}^{\infty} \int_{u=0}^{2\pi} f(u) \{\cos nu.\cos nx + \sin nu.\sin nx)\}\, du$$

$$= (1/2\pi) \int_{u=0}^{2\pi} f(u)\, du + (1/\pi) \sum_{n=1}^{\infty} [\{\int_{u=0}^{2\pi} f(u).\cos nu.\, du\} \cos nx$$

$$+ \{\int_{u=0}^{2\pi} f(u).\sin nu\,.du\} \sin nx]. \qquad (3.19b)$$

Note 3.2. Comparing the results given by Eqs. (3.19a) and (3.19b) with Eqs. (2.11) and (2.10) respectively, we note that the difference lies in the limits only.

3.6. The interval [0, 1]: Rewriting the Fourier series expansion given by Eq. (5.19a) for a function $F(y)$ defined in the interval $[0, 2\pi]$:

$$F(y) = (1/2\pi) \int_{t=0}^{2\pi} F(t)\, dt$$

$$+ (1/\pi) \sum_{n=1}^{\infty} \int_{t=0}^{2\pi} F(t) \cos \{n\,(t-y)\}dt, \qquad (3.20)$$

and applying the change of parameters $t = 2\pi u$ so that $dt = 2\pi\, du$ and $y = 2\pi x$, we derive

$$F(2\pi x) = \int_{u=0}^{1} F(2\pi u)\, du + 2 \sum_{n=1}^{\infty} \int_{u=0}^{1} F(2\pi u) \cos \{2n\pi\,(u-x)\}du.$$

Setting $F(2\pi x) = f(x)$ and $F(2\pi u) = f(u)$, above expansion reads as

$$f(x) = \int_{u=0}^{1} f(u)\, du + 2 \sum_{n=1}^{\infty} \int_{u=0}^{1} f(u) \cos \{2n\pi\,(u-x)\}\, du$$

$$= \int_{u=0}^{1} f(u)\, du + 2 \sum_{n=1}^{\infty} [\{\int_{u=0}^{1} f(u) \cos (2n\pi u).\, du\} \cos (2n\pi x)$$

$$+ \{ \int_{u=0}^{1} f(u) \sin (2n\pi u).du\}.\sin (2n\pi x)]. \qquad (3.21)$$

Particularly, the *cosine* (respectively *sine*) series expansions for a function defined in the interval [0, 1] can be found in analogy with Eq. (3.14) respectively Eq. (3.16):

$$f(x) = \int_{u=0}^{1} f(u)\, du$$

$$+ 2 \sum_{n=1}^{\infty} \{ \int_{u=0}^{1} f(u) \cos (n\pi u)\, du \}. \cos (n\pi x) \qquad (3.22)$$

(respectively)

$$f(x) = 2 \sum_{n=1}^{\infty} \{ \int_{u=0}^{1} f(u) \sin (n\pi u)\, du \}. \sin (n\pi x). \; // \qquad (3.23)$$

Example 3.6. Find the Fourier series expansion for the function $f(x) = e^{2x}$ defined on the interval [0, 1].

Solution. The Fourier coefficients in the expansion given by Eq. (3.21) are:

$$a_0 = \int_{x=0}^{1} e^{2x}\, dx = \left[e^{2x} / 2 \right]_{0}^{1} = (e^2 - 1) / 2,$$

$$a_n = 2 \int_{x=0}^{1} e^{2x}. \cos (2n\pi x)\, dx = \left[e^{2x} (\cos 2n\pi x + n\pi.\sin 2n\pi x) \right]_{0}^{1} / (n^2 \pi^2 + 1)$$

$$= (e^2. \cos 2n\pi - 1) / (n^2 \pi^2 + 1) = (e^2 - 1) / (n^2 \pi^2 + 1),$$

and

$$b_n = 2 \int_{x=0}^{1} e^{2x}. \sin (2n\pi x)\, dx = \left[e^{2x} (\sin 2n\pi x - n\pi.\cos 2n\pi x) \right]_{0}^{1} / (n^2 \pi^2 + 1)$$

$$= n\pi (- e^2.\cos 2n\pi + 1) / (n^2 \pi^2 + 1) = - n\pi (e^2 - 1) / (n^2 \pi^2 + 1),$$

where we have applied the integral formulae vide Eqs. (1.1.9) and (1.1.10) (cf. [17], Eqs. (11.2.38) and (11.2.39))

$$\left. \begin{array}{l} \int e^{ax}.\sin bx.\, dx = e^{ax}.(a.\sin bx - b.\cos bx) / (a^2 + b^2), \\[2mm] \int e^{ax}.\cos bx.\, dx = e^{ax}.(a.\cos bx + b.\sin bx) / (a^2 + b^2). \end{array} \right\} \qquad (3.24)$$

and

Thus, the desired Fourier series is

$$e^{2x} = (e^2 - 1)\left[1/2 + \sum_{n=1}^{\infty} \{\cos(2n\pi x) - n\pi.\sin(2n\pi x)\}/(n^2\pi^2 + 1)\right]. \quad (3.25)$$

3.7. The interval [a, b]: Applying the change of parameters

$$t = 2\pi(u-a)/(b-a) \implies dt = 2\pi\,du/(b-a) \text{ and } y = 2\pi(x-a)/(b-a),$$

the equation (3.20) reads as

$$F\{2\pi(x-a)/(b-a)\} = \left[\int_{u=a}^{b} F\{2\pi(u-a)/(b-a)\}\,du\right.$$

$$+ 2\sum_{n=1}^{\infty}\int_{u=a}^{b} F\{2\pi(u-a)/(b-a)\}\cos\{2n\pi(u-x)/(b-a)\}du]/(b-a)$$

or,
$$f(x) = \left[\int_{u=a}^{b} f(u)\,du\right.$$

$$+ 2\sum_{n=1}^{\infty}\int_{u=a}^{b} f(u)\cos\{2n\pi(u-x)/(b-a)\}du]/(b-a). \quad (3.26)$$

§ 4. Fourier series for piecewise defined functions

Let a function $f(x)$ be defined on some closed interval $[a, b]$ at points of the interval except for a finite number of points, say $x_1, x_2, \ldots, x_n$. It is called *piecewise continuous* on the interval $[a, b]$ if:

(i) it is continuous on each sub-intervals $(a, x_1), (x_1, x_2), \ldots, (x_{n-1}, x_n),$ (x_n, b);

(ii) it possesses a finite limit from the right at the (left) end $x = a$, a finite limit from the left at the (right) end $x = b$, and both left and right limits at each point x_i, $i = 1, 2, \ldots, n$.

We shall obtain Fourier series expansions for such functions in the following. The method is demonstrated by means of some examples.

Example 4.1. Find the Fourier series for the function defined on the interval $[-2, 2]$:

$$f(x) = 0 \text{ (when } -2 \le x < 0), \ 1 \text{ (when } 0 \le x < 1), \ 2 \text{ (when } 1 \le x \le 2).$$

Solution. A Fourier series expansion of a function defined over the interval $[-l, l]$ is obtained by Eq. (3.3) and the corresponding Fourier coefficients by Eq. (3.4). Thus, for $l = 2$, we have

$$a_0 = (1/4) \int_{-2}^{2} f(x)\, dx = (1/4) \{ \int_{-2}^{0} + \int_{0}^{1} + \int_{1}^{2} \} f(x)\, dx$$

$$= (1/4) \{ 0 + \int_{0}^{1} dx + 2 \int_{1}^{2} dx \} = (1/4) \{ [x]_0^1 + 2 [x]_1^2 \}$$

$$= (1/4)(1 + 2) = 3/4,$$

$$a_n = (1/2) \int_{-2}^{2} f(x) \cos(n\pi x/2)\, dx$$

$$= (1/2) \{ \int_{-2}^{0} + \int_{0}^{1} + \int_{1}^{2} \}.f(x) \cos(n\pi x/2)\, dx$$

$$= (1/2) \{ \int_{0}^{1} \cos(n\pi x/2)\, dx + 2 \int_{1}^{2} \cos(n\pi x/2)\, dx \}$$

$$= (1/n\pi) \{ [\sin(n\pi x/2)]_0^1 + 2 [\sin(n\pi x/2)]_1^2 \}$$

$$= (1/n\pi) [\sin(n\pi/2) + 2\{\sin n\pi - \sin(n\pi/2)\}] = - \{\sin(n\pi/2)\}/n\pi,$$

and

$$b_n = (1/2) \int_{-2}^{2} f(x) \sin(n\pi x/2)\, dx$$

$$= (1/2) \{ \int_{-2}^{0} + \int_{0}^{1} + \int_{1}^{2} \}.f(x) \sin(n\pi x/2)\, dx$$

$$= (1/2) \{ \int_{0}^{1} \sin(n\pi x/2)\, dx + 2 \int_{1}^{2} \sin(n\pi x/2)\, dx \}$$

$$= - (1/n\pi) \{ [\cos(n\pi x/2)]_0^1 + 2 [\cos(n\pi x/2)]_1^2 \}$$

$$= - (1/n\pi) [\cos(n\pi/2) - 1 + 2 \{\cos(n\pi) - \cos(n\pi/2)\}]$$

$$= (1/n\pi) \{ 1 + \cos(n\pi/2) - 2.\cos n\pi \}.$$

Thus, the desired Fourier series for above function is

$$f(x) = 3/4 + (1/n\pi). \sum_{n=1}^{\infty} [- \sin(n\pi/2).\cos(n\pi x/2)$$

$$+ \{ 1 + \cos(n\pi/2) - 2.\cos n\pi \}. \sin(n\pi x/2)]. //$$

Example 4.2. Find the Fourier series for the function defined by

$$f(x) = -1 \text{ (when } -1 \leq x < 0), \quad 1 \text{ (when } 0 \leq x \leq 1)$$

on the interval $[-1, 1]$.

Solution. The equation (3.4), for $l = 1$, determines the Fourier coefficients:

$$a_0 = (1/2) \int_{-1}^{1} f(x) \, dx = (1/2) \{ \int_{-1}^{0} + \int_{0}^{1} \} f(x) \, dx$$

$$= (1/2) \{ -\int_{-1}^{0} dx + \int_{0}^{1} dx \} = (1/2) \{ [x]_0^{-1} + [x]_0^{1} \} = (1/2)(-1 + 1) = 0,$$

$$a_n = \int_{-1}^{1} f(x) \cos(n\pi x) \, dx = \{ \int_{-1}^{0} + \int_{0}^{1} \} f(x) \cos(n\pi x) \, dx$$

$$= -\int_{-1}^{0} \cos(n\pi x) \, dx + \int_{0}^{1} \cos(n\pi x) \, dx$$

$$= (1/n\pi) \{ [\sin(n\pi x)]_0^{-1} + [\sin(n\pi x)]_0^{1} \} = 0,$$

and

$$b_n = \int_{-1}^{1} f(x). \sin(n\pi x) \, dx = \{ \int_{-1}^{0} + \int_{0}^{1} \} f(x). \sin(n\pi x) \, dx$$

$$= -\int_{-1}^{0} \sin(n\pi x) \, dx + \int_{0}^{1} \sin(n\pi x) \, dx$$

$$= \{ [\cos(n\pi x)]_{-1}^{0} - [\cos(n\pi x)]_0^{1} \}/n\pi = (1 - \cos n\pi - \cos n\pi + 1) / n\pi$$

$$= 2(1 - \cos n\pi) / n\pi = 0 \quad \text{(for even } n\text{)}, \qquad 4/n\pi \quad \text{(for odd } n\text{)}.$$

Thus, the desired Fourier series for above function is

$$f(x) = 2 \sum_{n=1}^{\infty} \{ (1 - \cos n\pi) / n\pi \} \sin(n\pi x)$$

$$= (4/\pi) \sum_{m=1}^{\infty} \sin \{ (2m-1) \pi x) \} / (2m-1), \quad n \text{ being odd, say } 2m-1. \, //$$

Example 4.3. Find the Fourier series expansion for the function defined by

$$f(x) = k x \text{ (when } 0 \leq x \leq l/2), \quad k(l-x) \text{ (when } l/2 \leq x \leq l)$$

on the interval $[0, l]$.

Solution. (i) The *cosine* series expansion of a function defined on the interval $[0, l]$ is given by the equation (3.14). Breaking the range of integration therein, we have

$$a_0 = (1/l)\int_0^l f(x)\,dx = (1/l)\{\int_0^{l/2} + \int_{l/2}^l \}f(x)\,dx$$

$$= (k/l)\{\int_0^{l/2} x\,dx + \int_{l/2}^l (l-x)\,dx\}$$

$$= (k/l)\{[x^2/2]_0^{l/2} + [lx - x^2/2]_{l/2}^l\}$$

$$= (k/l)(l^2/8 + l^2 - l^2/2 - l^2/2 + l^2/8) = kl/4,$$

and

$$a_n = (2/l)\int_0^l f(x)\cos(n\pi x/l)\,dx$$

$$= (2/l)\{\int_0^{l/2} + \int_{l/2}^l \}f(x)\cos(n\pi x/l)\,dx$$

$$= (2k/l)\{\int_0^{l/2} x.\cos(n\pi x/l)\,dx + \int_{l/2}^l (l-x)\cos(n\pi x/l)\,dx\}$$

$$= (2k/n\pi)\{[x\sin(n\pi x/l)]_0^{l/2} - \int_0^{l/2}\sin(n\pi x/l)\,dx$$

$$+ [(l-x)\sin(n\pi x/l)]_{l/2}^l + \int_{l/2}^l \sin(n\pi x/l)\,dx\}$$

$$= (2k/n\pi)\{(l/2)\sin(n\pi/2) + (l/n\pi)[\cos(n\pi x/l)]_0^{l/2}$$

$$- (l/2)\sin(n\pi/2) - (l/n\pi)[\cos(n\pi x/l)]_{l/2}^l\}$$

$$= (2kl/n^2\pi^2)\{\cos(n\pi/2) - 1 - \cos n\pi + \cos(n\pi/2)\}$$

$$= (2kl/n^2\pi^2)\{2\cos(n\pi/2) - 1 - \cos n\pi\}$$

$$= (2kl/n^2\pi^2)\{2\cos(n\pi/2)(1 - \cos(n\pi/2)\},$$

which vanishes for all odd values of n as well as for even values of $n/2$, i.e. when $n = 4, 8, 12, 16, \ldots$ But, for odd values of $n/2$, i.e. when $n = 2$,

6, 10, 14, ... the expression within the curly brackets is -4 determining the following values of a_n :

$$a_n = -8kl / n^2 \pi^2, \text{ for } n = 2, 6, 10, 14, ...$$

Accordingly, the desired Fourier series is

$$f(x) = kl/4 - (8kl/\pi^2) \ [\ \{\cos (2\pi x / l)\}/ 2^2 + \{\cos (6\pi x / l)\}/ 6^2$$

$$+ \{\cos (10\pi x / l)\}/ 10^2 + ... \ ].$$

(ii) On the other hand, the *sine* series expansion of a function defined on the interval $[0, l]$ is given by Eq. (3.16). Proceeding similarly, we evaluate

$$\int_0^l f(x). \sin (n\pi x / l) \, dx = \{\int_0^{l/2} + \int_{l/2}^l \} f(x) \sin (n\pi x / l) \, dx$$

$$= k \{\int_0^{l/2} x. \sin (n\pi x / l) \, dx + \int_{l/2}^l (l - x) \sin (n\pi x / l) \, dx\}$$

$$= (kl / n\pi) \ \{ \left[-x \cos (n\pi x / l)\right]_0^{l/2} + \int_0^{l/2} \cos (n\pi x / l) \, dx$$

$$- \left[(l - x) \cos (n\pi x / l)\right]_{l/2}^l - \int_{l/2}^l \cos (n\pi x / l) \, dx \}$$

$$= (kl / n\pi) \ \{ - (l/2) \cos (n\pi/2) + (l / n\pi)\left[\sin (n\pi x / l)\right]_0^{l/2}$$

$$+ (l/2) \cos (n\pi/2) - (l / n\pi)\left[\sin (n\pi x / l)\right]_{l/2}^l \}$$

$$= (kl^2 / n^2\pi^2) \ \{\sin (n\pi/2) - \sin n\pi + \sin (n\pi/2)\} = (2kl^2 / n^2\pi^2) \sin (n\pi/2),$$

which vanishes for even values of n. Accordingly, the expansion (3.16) reduces to

$$f(x) = (4kl/\pi^2) \sum_{n=1}^{\infty} \ \{\sin (n\pi/2). \sin (n\pi x / l)\} / n^2$$

$$= (4kl/\pi^2) \ [\{\sin (\pi x/l)\}/1^2 - \{\sin (3\pi x/l)\}/3^2 + \{\sin (5\pi x/l)\}/5^2 - ...\}. \ //$$

Example 4.4. Find the Fourier series expansion for the following function defined on the interval $[-2l, 2l]$:

$$f(x) = \begin{cases} l & \text{(when } -2l \leq x \leq -l \text{),} \\ -x & \text{(when } -l \leq x \leq 0 \text{),} \\ x & \text{(when } 0 \leq x \leq l \text{),} \\ l & \text{(when } l \leq x \leq 2l \text{).} \end{cases} \qquad (4.1)$$

Solution. Eq. (3.1) gives the Fourier series expansion of a function $F(y)$ on the interval $[-\pi, \pi]$. Applying the change of parameters as:

$$t = \pi u / 2l \implies dt = (\pi \, du) / 2l, \text{ and } \quad y = \pi x / 2l.$$

Eq. (3.1) assumes the form

$$F(\pi x / 2l) = (1/4l) \int_{-2l}^{2l} F(\pi u / 2l) \, du$$

$$+ (1/2l) \sum_{n=1}^{\infty} \int_{-2l}^{2l} F(\pi u / 2l) \cos \{n\pi (u-x)/2l\} \, du$$

Writing $F(\pi x / 2l)$ as $f(x)$, and similarly $F(\pi u/2l)$ as $f(u)$ above relation further alters as

$$f(x) = (1/4l) \int_{-2l}^{2l} f(u) \, du$$

$$+ (1/2l) \sum_{n=1}^{\infty} \int_{-2l}^{2l} f(u) \cos \{n\pi (u-x)/2l\} \, du. \qquad (4.2)$$

Since

$$\int_{-2l}^{2l} f(u) \, du = \left\{ \int_{-2l}^{-l} + \int_{-l}^{0} + \int_{0}^{l} + \int_{l}^{2l} \right\} f(u) \, du$$

$$= \left\{ l \int_{-2l}^{-l} - \int_{-l}^{0} u + \int_{0}^{l} u + l \int_{l}^{2l} \right\} du$$

$$= l(-l+2l) + l^2/2 + l^2/2 + l(2l-l) = 3l^2,$$

and

$$\int_{-2l}^{2l} f(u) \cos \{n\pi (u-x)/2l\} \, du$$

$$= \left\{ \int_{-2l}^{-l} + \int_{-l}^{0} + \int_{0}^{l} + \int_{l}^{2l} \right\} f(u) \cos \{n\pi (u-x)/2l\} \, du$$

$$= l \int_{-2l}^{-l} \cos \{n\pi (u-x)/2l\} \, du - \int_{-l}^{0} u. \cos \{n\pi (u-x)/2l\} \, du$$

$$+ \int_0^l u \cdot \cos \{ n\pi (u-x)/2l \} \, du + l \int_l^{2l} \cos \{ n\pi (u-x)/2l \} \, du$$

$$= (2l/n\pi) \left\{ l \left[\sin\{n\pi(u-x)/2l\} \right]_{-2l}^{-l} + \left[u\sin\{n\pi(u-x)/2l\} \right]_0^{-l} \right.$$

$$- \int_0^{-l} \sin \{ n\pi (u-x)/2l \} \, du + \left[u\sin\{n\pi(u-x)/2l\} \right]_0^l$$

$$- \int_0^l \sin \{ n\pi (u-x)/2l \} \, du + l \left. \left[\sin\{n\pi(u-x)/2l\} \right]_l^{2l} \right\}$$

$$= (2l^2/n\pi) \left(- \sin \{ n\pi (l+x)/2l \} + \sin \{ n\pi (2l+x)/2l \} \right.$$

$$+ \sin \{ n\pi (l+x)/2l \} + (2/n\pi) \left[\cos\{n\pi(u-x)/2l\} \right]_0^{-l}$$

$$+ \sin \{ n\pi (l-x)/2l \} + (2/n\pi) \left[\cos\{n\pi(u-x)/2l\} \right]_0^l$$

$$+ \sin \{ n\pi (2l-x)/2l \} - \sin \{ n\pi (l-x)/2l \} \left. \right)$$

$$= (2l^2/n\pi) \left(\sin \{ n\pi (2l+x)/2l \} + (2/n\pi) [\cos \{ n\pi (l+x)/2l \} \right.$$

$$- \cos (n\pi x/2l) + \{\cos \{ n\pi (l-x)/2l \} - \cos (n\pi x/2l)]$$

$$+ \sin \{ n\pi (2l-x)/2l \} \left. \right)$$

$$= (2l^2/n\pi) [\sin \{ n\pi (2l+x)/2l \} + \sin \{ n\pi (2l-x)/2l \}]$$

$$+ (2l/n\pi)^2 [\cos \{ n\pi (l+x)/2l \} + \cos \{ n\pi (l-x)/2l \} - 2 \cdot \cos (n\pi x/2l)]$$

$$= (4l^2/n\pi) \sin n\pi \cdot \cos (n\pi x/2l)$$

$$+ (2l/n\pi)^2 \{ 2 \cdot \cos (n\pi/2) \cdot \cos (n\pi x/2l) - 2 \cdot \cos (n\pi x/2l) \}$$

$$= (8l^2/n^2\pi^2) \{ \cos (n\pi/2) - 1 \} \cos (n\pi x/2l),$$

which is 0 when $n/2$ is even, $- (4l/n\pi)^2 \cdot \cos (n\pi x/2l)$ when $n/2$ is odd, and $- (8l^2/n^2\pi^2) \cdot \cos (n\pi x/2l)$ for odd values of n. Hence, Eq. (4.2) becomes

$$f(x) = (3l/4) + (4l/\pi^2) \sum_{n=1}^{\infty} [\{\cos(n\pi/2) - 1\} \cos(n\pi x/2l)] / n^2$$

$$= (3l/4) - (4l/\pi^2) \{\cos(\pi x/2l) + (2/2^2)\cos(2\pi x/2l) + (1/3^2)\cos(3\pi x/2l)$$

$$+ (1/5^2)\cos(5\pi x/2l) + (2/6^2)\cos(6\pi x/2l) + ...\}$$

$$= (3l/4) - (4l/\pi^2) [\{\cos(\pi x/2l) + (1/3^2)\cos(3\pi x/2l) + (1/5^2)\cos(5\pi x/2l)$$

$$+ ...\}$$

$$+ 2\{(1/2^2)\cos(\pi x/l) + (1/6^2)\cos(3\pi x/l) + (1/10^2)\cos(5\pi x/l) + ...\}].//$$

Example 4.5. Find a series of *sines* of multiples of *x* which represent the function

$$f(x) = \begin{cases} x/2 & \text{(when } 0 \leq x \leq \alpha), \\ \alpha/2 & \text{(when } \alpha \leq x \leq \pi - \alpha), \\ (\pi - x)/2 & \text{(when } \pi - \alpha \leq x \leq \pi), \end{cases}$$

on the interval $[0, \pi]$.

Solution. As in Example 3.3, in order to have a *sine* series expansion we construct a new (odd) function $g(x)$ on the extended interval $[-\pi, \pi]$. Thus, in view of Eq. (2.13), the Fourier coefficients are: $a_0 = a_n = 0$, and

$$b_n = (2/\pi) . \int_0^{\pi} f(x) . \sin nx \, dx$$

$$= (2/\pi) . \{ \int_0^{\alpha} + \int_{\alpha}^{\pi - \alpha} + \int_{\pi - \alpha}^{\pi} \} f(x) . \sin nx \, dx$$

$$= (1/\pi) . \{ \int_0^{\alpha} x . \sin nx \, dx + \alpha \int_{\alpha}^{\pi - \alpha} \sin nx \, dx + \int_{\pi - \alpha}^{\pi} (\pi - x) . \sin nx \, dx$$

$$= (1/n\pi) \{ \left[- x . \cos nx \right]_0^{\alpha} + \int_0^{\alpha} \cos nx \, dx - \alpha \left[\cos nx \right]_{\alpha}^{\pi - \alpha}$$

$$- \left[(\pi - x) \cos nx \right]_{\pi - \alpha}^{\pi} - \int_{\pi - \alpha}^{\pi} \cos nx \, dx \}$$

$$= (1/n\pi) (- \alpha \cos n\alpha + (1/n) \left[\sin nx \right]_0^{\alpha} - \alpha \{ \cos n (\pi - \alpha) - \cos n\alpha \}$$

$$+ \alpha . \cos n (\pi - \alpha) - (1/n) \left[\sin nx \right]_{\pi - \alpha}^{\pi})$$

$$= (1/n^2 \pi) \{ \sin n\alpha + \sin n (\pi - \alpha) \}$$

$$= (2/n^2 \pi) \sin (n\pi /2) \cos n (\pi/2 - \alpha) = 0, \text{ if } n \text{ is even.}$$

Therefore, Eq. (2.10b) assumes the form

$$f(x) = (2/\pi) \sum_{n=1}^{\infty} \{ \sin (n\pi/2). \cos n (\pi/2 - \alpha). \sin nx \} / n^2$$

$$= (2/\pi)\{(\sin \alpha. \sin x)/1^2 + (\sin 3\alpha.\sin 3x)/3^2 + (\sin 5\alpha. \sin 5x)/5^2 + ...\}.//$$

CHAPTER 9

STATISTICAL TECHNIQUES

§ 1. Introduction

Statistics is basically a decision making science dealing with collections, organizations, analyzing and interpreting data. While applying statistics to a scientific, social or industrial problem it is conventional to begin with a *statistical population* or *statistical model* process. Population may be of diverse nature, such as 'people living in a country', 'every atom comprising of a crystal'. Two main statistical methods employed in data analysis are:

(i) Descriptive statistics – summarizing data from a sample by using mean and standard deviation;

and

(ii) Inferential statistics – drawing conclusions from data subject to random variation.

The descriptive statistics is mainly concerned with four basic characteristics: average, dispersion, skewness and kurtosis. On the other hand, *inferences* on mathematical statistics are made under Probability Theory that deals with the analysis of random phenomena. A standard statistical procedure involves "the tests of the relationship between two data sets". Rejection or disapproval of the null hypothesis is based on statistical tests. Multiple problems have come up to be associated with this framework.

§ 2. Moment

A distribution: discrete or continuous may have more characteristics (like mean, median, and mode) than the usual ones. It is not possible to interpret all the characteristics of a distribution in physical terms such as symmetry and dispersion. However, there are mathematical methods which distinguish one distribution from the other. Method of *moments* is one of these.

2.1. The rth moment about origin, denoted by μ_r', for a distribution

$$\begin{pmatrix} x_i \\ f_i \end{pmatrix} = \begin{pmatrix} x_1 & x_2 & \cdot & \cdot & x_n \\ f_1 & f_2 & \cdot & \cdot & f_n \end{pmatrix} \qquad (2.1)$$

is defined by

$$\mu_r' = (1/N). \sum_{i=1}^{n} f_i x_i^r, \qquad \text{where} \qquad N = \sum_{i=1}^{n} f_i. \qquad (2.2)$$

Particularly, for $r = 0$,

$$\mu_0' = (1/N). \sum_{i=1}^{n} f_i x_i^0 = (1/N). \sum_{i=1}^{n} f_i = 1, \qquad (2.3)$$

by Eq. (2.2). The first and second moments about origin are

$$\mu_1' \equiv \bar{x} = (1/N). \sum_{i=1}^{n} f_i x_i, \qquad \mu_2' = (1/N). \sum_{i=1}^{n} f_i x_i^2 .$$
$$(2.4)$$

It may be noted that μ_1' is the *Arithmetic Mean* of the distribution while μ_2' is called the *mean square*.

2.2. Moment about mean value

Such *moments* are of special importance in the study of various characteristics of a distribution. Denoting the rth moment of distribution about mean by μ_n it is defined by

$$\mu_r = (1/N). \sum_{i=1}^{n} f_i (x_i - \bar{x})^r. \qquad (2.5)$$

Thus, in particular, in every distribution

$$\mu_1 = (1/N). \sum_{i=1}^{n} f_i x_i - (\bar{x}/N). \sum_{i=1}^{n} f_i = 0, \qquad (2.6)$$

by Eqs. (2.2) and (2.4). Also,

$$\mu_2 = (1/N). \sum_{i=1}^{n} f_i (x_i - \bar{x})^2 = \sigma^2, \qquad (2.7)$$

is the variance in every distribution. Further,

$$\mu_3 = (1/N). \sum_{i=1}^{n} f_i (x_i - \bar{x})^3, \qquad (2.8)$$

and

$$\mu_4 = (1/N). \sum_{i=1}^{n} f_i (x_i - \bar{x})^4, \qquad (2.9)$$

which are used in the study of *skewness* and *kurtosis* dealing with the symmetry and peakedness of the distribution.

2.3. Moment about any value

Similarly, the moments about an arbitrary value a are defined by

$$\mu_r'(a) = (1/N). \sum_{i=1}^{n} f_i (x_i - a)^r. \tag{2.10}$$

When $a = \bar{x}$, it coincides with μ_r for Eq. (2.5). Expanding $(x_i - \bar{x})^r$ by binomial expansion in Eq. (2.5) we, thus, have

$$\mu_r = (1/N). \sum_{i=1}^{n} f_i \{x_i^r - {}^rC_1 \bar{x} x_i^{r-1} + {}^rC_2 \bar{x}^2 x_i^{r-2} - .... + (-1)^r \bar{x}^r\}$$

$$= \mu_r' - {}^rC_1 \bar{x} \mu_{r-1}' + {}^rC_2 \bar{x}^2 \mu_{r-2}' - .... + (-1)^r \bar{x}^r, \tag{2.11}$$

by Eq. (2.2). Giving different integral values to r, above equation yields

$$\mu_1 = \mu_1' - \bar{x} \mu_0' = \bar{x} - \bar{x} = 0, \tag{2.12}$$

$$\mu_2 = \mu_2' - 2\bar{x}\mu_1' + \bar{x}^2 \mu_0' = \mu_2' - 2\bar{x}^2 + \bar{x}^2$$

$$= \mu_2' - \bar{x}^2 = \mu_2' - (\mu_1')^2, \tag{2.13}$$

$$\mu_3 = \mu_3' - 3\bar{x}\mu_2' + 3\bar{x}^2 \mu_1' - \bar{x}^3 = \mu_3' - 3\bar{x}\mu_2' + 2\bar{x}^3$$

$$= \mu_3' - 3\mu_1'\mu_2' + 2(\mu_1')^3, \tag{2.14}$$

$$\mu_4 = \mu_4' - 4\bar{x}\mu_3' + 6\bar{x}^2 \mu_2' - 4\bar{x}^3\mu_1' + \bar{x}^4$$

$$= \mu_4' - 4\mu_1'\mu_3' + 6(\mu_1')^2 \mu_2' - 3(\mu_1')^4, \tag{2.15}$$

etc. where substitutions are made from Eqs. (2.3) and (2.4).

2.4. Change of origin and unit

Shifting the origin to the point a and taking unit b, there results a transformation of the coordinates:

$$x_i = a + b\,u_i, \qquad \text{and similarly} \qquad \bar{x} = a + b\,\bar{u},$$

$$\Rightarrow$$

$$x_i - \bar{x} = b\,(u_i - \bar{u}\,).$$

Accordingly, Eq. (2.5) determines

$$\mu_r = (b^r/N).\sum_{i=1}^{n} f_i\,(u_i - \bar{u})^r = b^r.(\text{moment of } u \text{ about } \bar{u})$$

concluding:

Theorem 2.1. Moments about the mean remains unaltered under the change of origin.

Example 2.1. Given the first four moments of a distribution about a value as -4, 17, -2 and 301 respectively, find their values about the mean.

Solution. Let the origin be shifted to the given value. Moments after this shift are computed in Sub-section 2.3 above. Thus, as per hypothesis, we have

$$\mu_1' = -4, \quad \mu_2' = 17, \quad \mu_3' = -2, \text{ and } \quad \mu_4' = 301.$$

Putting for these values, Eqs. (2.10) – (2.13) determine

$$\mu_1 = 0, \ \mu_2 = 17 - (-4)^2 = 1, \ \mu_3 = -2 - 3.(-4).17 + 2\,(-4)^3 = 74,$$

and

$$\mu_4 = 301 - 4\,(-4)\,(-2) + 6.\,(-4)^2.\,17 - 3.\,(-4)^4 = 1133. \ //$$

Example 2.2. Show that the second moment is minimum about the mean value.

Solution. The second moment about mean value is given by Eq. (2.12) that yields

$$\mu_2' = \mu_2 + (\mu_1')^2.$$

The second term on RHS being square (of a real value) in above equation is never negative. It attains the minimum value zero leaving the equation

$$\mu_2' = \mu_2 .$$

Thus, the minimum value of μ_2' is the moment about mean value. //

2.5. Moment of a grouped frequency distribution

In order to reduce the grouped frequency distribution to ungrouped distribution, each class of the distribution is replaced by the mid-value and compute the moments in a similar way. Obviously, these are the approximate values of the moments.

For a continuous distribution the rth moment about origin is given by

$$\mu_r' = \int_a^b x^r . \varphi(x) \, dx, \tag{2.16}$$

where the frequency function $\varphi(x)$ represents the distribution in the interval $a \le x \le b$:

$$y = \varphi(x). \tag{2.17}$$

The rth moment about the mean value $\bar{x}$ is analogously given by

$$\mu_r = \int_a^b (x - \bar{x})^r . \varphi(x) \, dx, \tag{2.18}$$

Writing the binomial expansion of $(x - \bar{x})^r$, and following the procedure of Sub-section 2.3, one can similarly derive

$$\mu_r = \mu_r' - {}^rC_1 \mu_1' \mu_{r-1}' + {}^rC_2 (\mu_1')^2 \mu_{r-2}' - \dots + (-1)^r (\mu_1')^r, \tag{2.19}$$

§ 3. Moment generating function (m.g.f.)

3.1. Continuous distribution

For a continuous distribution given by Eq. (2.17), the function defined by

$$M(t) = \int_a^b e^{tx} . \varphi(x) \, dx, \tag{3.1}$$

is called the *moment generating function* (m.g.f.) of the distribution about the origin. Expanding the exponential function e^{tx} in powers of t x, above function may be re-written as

$$M(t) = \int_a^b \{1 + tx + (tx)^2/2! + \ldots + (tx)^r/r! + \ldots\}.\varphi(x).dx$$

$$= \int_a^b \varphi(x).dx + t.\int_a^b x.\varphi(x).dx + (t^2/2!).\int_a^b x^2.\varphi(x).dx$$

$$+ (t^r/r!).\int_a^b x^r.\varphi(x).dx + \ldots \qquad (3.2)$$

Supposing that the RHS is term wise integrable, comparing it with the distribution in Sub-section 2.1, and writing

$$\mu_r' = \int_a^b x^r.\varphi(x).dx, \qquad (3.3)$$

$$\mu_0' = \int_a^b \varphi(x).dx = 1, \ \mu_1' = \int_a^b x.\varphi(x).dx, \ \mu_2' = \int_a^b x^2.\varphi(x).dx, \quad (3.4)$$

etc. Hence, Eq. (3.2) may be re-written as

$$M(t) = 1 + t.\mu_1' + (t^2/2!).\mu_2' + \ldots + (t^r/r!).\mu_r' + \ldots \qquad (3.5)$$

Note 3.1. In analogy with Eq. (2.10), m.g.f. of the continuous distribution in Eq. (2.17) about an arbitrary value, say c, may be defined by

$$M(t, c) = \int_a^b e^{t(x-c)}.\varphi(x).dx, \qquad (3.6)$$

which may be seen as a multiple of its counterpart about the origin:

$$M(t, c) = e^{-tc}.\int_a^b e^{tx}.\varphi(x).dx = e^{-tc}.M(t), \qquad (3.7)$$

by Eq. (3.1).

3.2. Ungrouped distribution

For an ungrouped distribution given by Eq. (2.1), the m.g.f. about origin is analogously defined by

$$M(t) = (1/N).\int_a^b f_i.\exp(tx_i).dx, \qquad (3.8)$$

Example 3.1. For a continuous distribution with relative frequency function

$$\varphi(x) = 3x.(2-x)/4, \quad 0 \le x \le 2,$$

show that the mean and variance of the distribution are 1 and 1/5 respectively.

Solution. (i) The arithmetic mean is obtained from Eqs. (2.4) and (2.16):

$$\mu_1' = \int_0^2 x.\, \varphi\,(x).\, dx = (3/4).\int_0^2 (2x^2 - x^3).\, dx$$

$$= \left[x^3/2 - 3x^4/16 \right]_0^2 = 1.$$

(iii) Also, Eqs. (2.4) and (2.16) yield

$$\mu_2' = \int_0^2 x^2.\, \varphi\,(x).\, dx = (3/4).\int_0^2 (2x^3 - x^4).\, dx$$

$$= \left[3x^4/8 - 3x^5/20 \right]_0^2 = 1/5.\; //$$

§ 4. Skewness

Skewness is related to the symmetric properties of a distribution. A distribution is either symmetrical or non-symmetrical about a central value. A non-symmetrical distribution is often called *skew distribution*. It is of two types:

(i) positively skew: when the frequency curve (or polygon) of distribution stretches longer towards right of the central value;

(ii) negatively skew: when the frequency curve (or polygon) of distribution stretches longer towards left of the central value.

For an ungrouped distribution in Eq. (2.1), the measure of skewness depends upon the third moment μ_3 about mean given by Eq. (2.8), which vanishes for a symmetric distribution. It is positive for a positively skew distribution as the variable x_i (being in the right of the mean value $\bar{x}$) goes higher than $\bar{x}$. On the other hand, μ_3 becomes negative for a negatively skew distribution as x_i stretches towards left of the mean value $\bar{x}$.

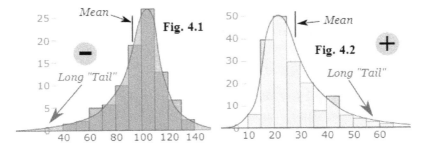

4.1. Coefficient of skewness

The measure of skewness independent of the unit of measurement is called the *coefficient of skewness* and it is denoted by γ_1. It is obtained by dividing μ_3 by σ^3, where σ is the standard deviation defined by

$$\sigma = \sqrt{\sum_{i=1}^{n} f_i (x_i - \bar{x})^2 / N}. \tag{4.1}$$

Thus, we have

$$\gamma_1 = \mu_3 / \sigma^3. \tag{4.2}$$

At times, the square of γ_1 (denoted as β_1 is also used as alternate coefficient of skewness. Thus, for a symmetrical distribution, $\gamma_1 = 0$; while it is positive (respectively negative) for a positively (resp. negatively) skew distribution.

§ 5. Kurtosis

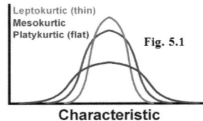

Fig. 5.1

Leptokurtic (thin)
Mesokurtic
Platykurtic (flat)

Characteristic

We consider three frequency curves possessing same mean, variance and skewness (Fig. 5.1). However, their graphs still differ: one of them has sharpest, other lowest and the remaining with middle hump. Such characteristic of the distribution is known as *kurtosis*. It measures the peakedness of the curve.

5.1. Measure of kurtosis

Kurtosis of a distribution is measured by means of the fourth moment μ_4 of the distribution about the mean value which is given by Eq. (2.9). The ratio β_2 of μ_4 and σ^4 :

$$\beta_2 = \mu_4 / \sigma^4 \tag{5.1}$$

measures the absolute value of kurtosis. It is usually compared with the kurtosis of the normal distribution given by

$$y = \exp(-x^2 / 2\sigma^2) / \sigma \sqrt{(2\pi)}, \quad -\infty \leq x \leq \infty. \tag{5.2}$$

Theorem 5.1. The kurtosis of above normal distribution is 3.

Proof. Its mean, given by Eqs. (2.4) and (3.4), is

$$\mu'_1 = (1/\sigma\sqrt{2\pi}).\int_{-\infty}^{\infty} x.\exp(-x^2/2\sigma^2).dx = 0, \qquad (5.3)$$

by Eq. (1.17). Also, from Eq. (3.3), we obtain

$$\mu'_2 = (1/\sigma\sqrt{2\pi}).\int_{-\infty}^{\infty} x^2.\exp(-x^2/2\sigma^2).dx$$

$$= (2/\sigma\sqrt{2\pi}).\int_{0}^{\infty} x^2.\exp(-x^2/2\sigma^2).dx,$$

and $\qquad \mu'_4 = (1/\sigma\sqrt{2\pi}).\int_{-\infty}^{\infty} x^4.\exp(-x^2/2\sigma^2).dx$

$$= (2/\sigma\sqrt{2\pi}).\int_{0}^{\infty} x^4.\exp(-x^2/2\sigma^2).dx.$$

Setting

$$x^2 = 2\sigma^2 t \quad\Rightarrow\quad x.dx = \sigma^2.dt \quad \text{and} \quad x = \sigma\surd(2t), \qquad (5.4)$$

above integrals reduce to

$$\mu'_2 = (2\sigma^2/\sqrt{\pi}).\int_{0}^{\infty} t^{1/2}.e^{-t}.dt = (2\sigma^2/\sqrt{\pi}).\Gamma(3/2) = \sigma^2, \qquad (5.5)$$

and

$$\mu'_4 = (4\sigma^4/\sqrt{\pi}).\int_{0}^{\infty} t^{3/2}.e^{-t}.dt = (4\sigma^4/\sqrt{\pi}).\Gamma(5/2) = 3\sigma^4. \quad (5.6)$$

Hence, by Eq. (5.1), $\beta_2 = 3$. //

Note 5.1. For vanishing mean of above normal distribution (cf. Eq. (5.3)), the moments considered in Sub-sections 2.1 and 2.2 are same.

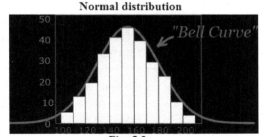

Fig. 5.2

Above theorem helps to consider another measure

$$\gamma_2 = \beta_2 - 3, \qquad (5.7)$$

called the *coefficient of excess*. When $\beta_2 > 3$ making γ_2 positive, the distribution is called *leptokurtic*. On the other hand, if $\beta_2 < 3$ so that γ_2 is negative, it is called *platykurtic* distribution. $\beta_2 = 3 \Rightarrow \gamma_2 = 0$ determines the middle path called *mesokurtic* distribution.

Note 5.2. Leptokurtic distributions have sharp peak at the mode while the peak at the mode will be flatter in case of platykurtic distributions.

§ 6. Correlation and regression

These are two diverse analysis based on a multivariate distribution involving multiple variables. *Correlation* is the analysis informing about an association or absence of a relationship between two variables, say x and y. On the other end, *regression* analysis, predicts a value of the dependent variable in terms of a known value of the independent variable. The following table depicts the main diverse characteristics of correlation and regression.

Characteristics	Correlation	Regression
Explanation	It is a statistical measure determining a relationship or association of two variables.	It describes how an independent variable is numerically related to the dependent variable.
Usage	It represents a linear relationship between the two variables.	It fits a best line and estimates one variable on the basis of another variable.
Dependent and independent variables	No difference.	Both variables are different.
Indication	Correlation coefficient indicates the extent to which two variables differ together.	It indicates the impact of a unit change in the known variable x on the estimated variable y.
Objective	To find a numerical value expressing the relationship between variables.	To estimate values of random variable on the basis of the values of fixed variable.

Here we discuss the distributions involving two or more variables. For instance, in a class of students, let us consider the height and weight of each student in the class. Thus, for each student, there are two measurable characters. In other words, changes in one variable are related to changes in other variable. This simultaneous variation is called *correla-*

tion. Denoting these variables as x and y we consider their pairs:

$$(x_1, y_1), \quad (x_2, y_2), \quad \ldots, \quad (x_i, y_i), \quad \ldots$$

It is probable that more than one pair may have same values of the variables. The number of times a pair occurs in the distribution is called the *frequency* of the distribution. Collection of pairs (x, y) with their frequencies is called the bivariate frequency distribution of x and y. The graph of the distribution representing a relationship between two characteristics may be a scattered diagram or stereogram.

Consider the pairs $(9, 23)$, $(16, 25)$, $(17, 35)$, $(20, 29)$, etc. and plot these as points on a 2-dimensional rectangular plane determined by x- and y-coordinate axes. The graph of these points may be a scattered one. On the other hand, a stereogram may look like a *histogram*. This representation gives a rough idea of relationship amongst x and y variables. Given a data, the curve of best fit is considered in Chapter 5, Example 2.1. The Eq. (5.2.8) representing the curve (indeed, a straight line) of best fit is called the *line of regression* of y on x and the coefficient b is called the *coefficient of regression* of y on x. In geometry, b is known as the slope of the straight line.

Putting the values of a and b obtainable from Eqs. (5.2.9) and (5.2.10) in Eq. (5.2.8) let the equation alters as

$$y - \bar{y} = b(x - \bar{x}), \tag{6.1}$$

where $\bar{x}$ (respectively $\bar{y}$) is the mean of values of x (resp. y):

$$\bar{x} = (x_1 + x_2 + \ldots + x_n) / n, \quad \text{and} \quad \bar{y} = (y_1 + y_2 + \ldots + y_n) / n. \tag{6.2}$$

On the other hand, to obtain a line of best fit to the values of x we consider

$$x = a' + b' y, \tag{6.3}$$

and proceed similarly as in Example 2.1, Chapter 5 to compute the values of coeffiencts a' and b'. Eq. represents the *line of regression* of x on y and the coefficient b' is called the *coefficient of regression* of x on y. Employing the mean values $\bar{x}$ and $\bar{y}$, let the Eq. (6.1) be analogously written in alternate form:

$$x - \bar{x} = b'(y - \bar{y}). \qquad (6.4)$$

Note 6.1. Two lines of regression represented by Eqs. (5.2.8) and (6.1) intersect each other at the mean value $(\bar{x}, \bar{y})$.

6.1. Coefficient of correlation

The coefficients b and b' give rise to another important coefficient γ defined by

$$\gamma = \sqrt{(b.\, b')}, \qquad (6.5)$$

called the *coefficient of correlation* or more precisely *coefficient of linear correlation* between the variables x and y.

Example 6.1. Two regression equations of the variables are x and y are

$$x = 19{\cdot}13 - 0{\cdot}87\, y, \quad \text{and} \quad y = 11{\cdot}64 - 0{\cdot}50\, x. \qquad (6.6)$$

Compute means $\bar{x}, \bar{y}$ and correlation coefficient between x and y.

Solution. (i) Since the mean values $\bar{x}, \bar{y}$ lie on the regression lines hence they also satisfy both of above equations:

$$\bar{x} = 19{\cdot}13 - 0{\cdot}87\, \bar{y}, \quad \text{and} \quad \bar{y} = 11{\cdot}64 - 0{\cdot}50\, \bar{x}.$$

Solving these simultaneous linear equations for $\bar{x}$ and $\bar{y}$ we get

$$\bar{x} = 15{\cdot}79 \quad \text{and} \quad \bar{y} = 3{\cdot}74.$$

(ii) The regression coefficients of x on y and vice-versa are found by Eqs. (6.6):

$$b' = -0{\cdot}87 \quad \text{and} \quad b = -0{\cdot}50;$$

which together with Eq. (6.2) determine the other coefficient

$$\gamma = \sqrt{(b.\, b')} = \sqrt{(0{\cdot}435)} = -0{\cdot}66.$$

Negative sign is accounted for b and b' both being negative. //

§ 7. Polynomial regression

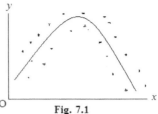

A linear relationship between two characteristics of bivariate distribution are discussed in the previous Section. For distributions as in Fig. 7.1, instead of a linear relationship there exists a curvilinear relationship:

Fig. 7.1

$$y = a_0 + a_1 x + a_2 x^2 + \dots + a_k x^k, \qquad (7.1)$$

where the index k is chosen as per nature of the distribution.

Now we fit the polynomial in accordance with the principle of least squares in the bivariate distribution

$$\begin{pmatrix} (x_i, y_i) \\ f_i \end{pmatrix} = \begin{pmatrix} (x_1, y_1) & (x_2, y_2) & \cdot & \cdot & (x_n, y_n) \\ f_1 & f_2 & \cdot & \cdot & f_n \end{pmatrix} \qquad (7.2)$$

As in Chapter 5, § 2, the sum of squares of deviations of actual values of y_i's from their estimated values η_i's is

$$S = \sum_{i=1}^{n} f_i (y_i - \eta_i)^2 = \sum_{i=1}^{n} f_i \{y_i - (a_0 + a_1 x_i + a_2 x_i^2 + \dots + a_k x_i^k)\}^2.$$

In order to have the least value of S there must hold the normal equations:

$$\partial S / \partial a_0 = \partial S / \partial a_1 = \partial S / \partial a_2 = \dots = \partial S / \partial a_k = 0,$$

yielding

$$\left. \begin{array}{l} \displaystyle\sum_{i=1}^{n} f_i \{y_i - (a_0 + a_1 x_i + a_2 x_i^2 + \dots + a_k x_i^k)\} = 0, \\[4mm] \displaystyle\sum_{i=1}^{n} f_i x_i \{y_i - (a_0 + a_1 x_i + a_2 x_i^2 + \dots + a_k x_i^k)\} = 0, \\[4mm] \displaystyle\sum_{i=1}^{n} f_i x_i^2 \{y_i - (a_0 + a_1 x_i + a_2 x_i^2 + \dots + a_k x_i^k)\} = 0, \\[4mm] \dots\dots\dots\dots\dots\dots\dots\dots\dots\dots\dots\dots\dots\dots\dots \\[4mm] \displaystyle\sum_{i=1}^{n} f_i x_i^k \{y_i - (a_0 + a_1 x_i + a_2 x_i^2 + \dots + a_k x_i^k)\} = 0. \end{array} \right\} \qquad (7.3)$$

These are $k + 1$ linear equations determining k unknown coefficients a_0, $a_1, a_2, \ldots, a_k$. Putting for these values of the coefficients a_k's in Eq. (7.1) we get the polynomial curve for regression of variable y on x.

Note 7.1. As a particular case, for $k = 2$, such polynomial regression curve has been considered in § 2, Chapter 5, where Eqs. (2.5) – (2.7) represent the concerned normal equations.

Analogously, one may also derive the polynomial curve for regression of variable x on y :

$$x = b_o + b_1 y + b_2 y^2 + \ldots + b_k y^k , \qquad (7.4)$$

where the coefficients b_k's shall be computed from $k + 1$ normal equations analogous to Eqs. (7.3).

Example 7.1. The amount of profit y earned by a company in x^{th} year of its establishment are given by

$$\begin{pmatrix} x \\ y \end{pmatrix} = \begin{pmatrix} 1 & 2 & 3 & 4 & 5 \\ 1250 & 1400 & 1650 & 1950 & 2300 \end{pmatrix}.$$

Show that the parabolic regression of y on x is represented by

$$y = 1140{\cdot}05 + 72{\cdot}1\, x + 32{\cdot}15\, x^2.$$

Solution. Values of x are symmetrical about its value 3, hence shifting the origin to 3 so that $x = u + 3$; and changing the units of y to v according to

$$y = 1650 + 50\, v , \qquad (7.5)$$

the distribution (of profits in the years) gets transformed as

$$\begin{pmatrix} u \\ v \end{pmatrix} = \begin{pmatrix} -2 & -1 & 0 & 1 & 2 \\ -8 & -5 & 0 & 6 & 13 \end{pmatrix} \qquad (7.6)$$

Let the parabolic regression of v on u be of the form

$$v = a + b\, u + c\, u^2. \qquad (7.7)$$

The normal equations (5.2.5) - (5.2.7) for above regression curve are

$$\Sigma v = 5a + b\,\Sigma u + c\,\Sigma u^2,$$

$$\Sigma uv = a\,\Sigma u + b\,\Sigma u^2 + c\,\Sigma u^3, \qquad (7.8)$$

and

$$\Sigma u^2 v = a\,\Sigma u^2 + b\,\Sigma u^3 + c\,\Sigma u^4.$$

Thus, computing the following values for the transformed distribution:

u	v	u^2	u^3	u^4	uv	$u^2 v$
-2	-8	4	-8	16	16	-32
-1	-5	1	-1	1	5	-5
0	0	0	0	0	0	0
1	6	1	1	1	6	6
2	13	4	8	16	26	52
$\Sigma u = 0$	$\Sigma v = 6$	Σu^2 $= 10$	Σu^3 $= 0$	Σu^4 $= 34$	Σuv $= 53$	$\Sigma u^2 v$ $= 21$

for which Eqs. (7.8) reduce to

$$5a + 10\,c = 6, \qquad 10\,b = 53, \qquad 10\,a + 34\,c = 21.$$

Solving these simultaneous linear equations we get $a = -0{\cdot}086$, $b = 5{\cdot}3$ and $c = 0{\cdot}643$. Putting for these values of a, b, c in Eq. (7.7), we get the regression curve

$$v = -0{\cdot}086 + 5{\cdot}3\,u + 0{\cdot}643\,u^2.$$

Finally, changing back to variables x and y as per Eq. (7.5) and $u = x - 3$, above equation assumes the desired form. //

§ 8. Theoretical distributions

Instead of actual observations, if we start with certain assumption and derive a distribution mathematically, then such distributions are called *theoretical distributions*. These distributions play an important role in statistics. Three different types of such distributions are considered here:

(i) Binomial distribution: a frequency distribution of the possible number of successful outcomes in a given number of trials in each of which there is the *same probability* of success. It was discovered by James Bernoulli (1654 - 1705 A.D.);

(ii) Poisson distribution: It is the *discrete probability* distribution of the number of events occurring in a given time period. It was discovered by French mathematician Siméon Denis Poisson in 1837 A.D.; and

(iii) Normal distribution: A function representing the distribution of many random variables as a symmetrical bell-shaped graph. It was discovered by Abraham de Moivre (26.5.1667 - 27.11.1754) in 1733. Later, in the end of 18th century, Pierre - Simon Laplace and Johann Carl Friedrich Gauss derived the normal distribution again independently.

8.1. Binomial distribution

In our day-to-day experience we come across problems such as tossing of a coin under similar conditions. We know that the outcome need not be the same in different tossings. In fact, there are innumerable causes influencing the result. Assuming n exhaustive, mutually exclusive and equally likely cases of which, say m ($< n$) are favourable to an event E, then $n - m$ cases are obviously unfavourable to the event. Thus, the probability for success of event E is p (E) $= m/n$ and that of failure of the event is q (E) $= 1 - m/n = (n - m) / n$. Evidently, these probabilities are connected by

$$p + q = 1. \tag{8.1}$$

Considering n trials of an experiment, the probability P_r for getting exactly r successes and $n - r$ failures of an event is

$$P_r = {}^n C_r . p^r . q^{n-r}. \tag{8.2}$$

Hence, the frequency f of the variable r (running over the values 0, 1, 2, ..., n) in N number of experiments is $N . {}^n C_r . p^r . q^{n-r}$. Representing it as a distribution

$$\begin{pmatrix} 0 & 1 & 2 & .. & r & .. & n \\ Nq^n & N {}^n C_1 p q^{n-1} & N {}^n C_2 p^2 q^{n-2} & .. & N {}^n C_r p^r q^{n-r} & .. & Np^n \end{pmatrix}, \tag{8.3}$$

we note that the frequencies are the successive terms in the binomial expansion of $N (q + p)^n$. Because of this characteristic, it is called a *binomial distribution* and the variable r as a binomial variable.

The moment generating function of the distribution about the origin may be found by Eq. (3.1):

$$M(t) = (1/N).\sum_{i=1}^{n} N.{}^nC_r p^r q^{n-r} e^{tr} = \sum_{i=1}^{n} {}^nC_r (pe')^r q^{n-r} = (q + pe')^n.$$

(8.4)

Therefore, cumulant generating function is

$$\psi(t) = \ln M(t) = n.\ln(q + p.\,e^t) = n.\ln[q + p.\{1 + t + t^2/2! + \dots \}]$$

$$= n.\ln\{1 + p.(t + t^2/2! + \dots)\}, \qquad \text{by Eq. (8.1).}$$

Applying expansion of logarithmic function we get

$$\psi(t) = n.\{p.(t + t^2/2! + \dots) - p^2.(t + t^2/2! + \dots)^2/2$$

$$+ p^3.(t + t^2/2! + \dots)^3/3 - p^4.(t + t^2/2! + \dots)^4/4 + \dots\}, \text{ for small values of } t$$

$$= n.\{pt + (p - p^2).\,t^2/2! + (p - 3p^2 + 2p^3).\,t^3/3!$$

$$+ (p - 7p^2 + 12p^3 - 6p^4).t^4/4! + \dots\}.$$

Therefore, we have

$$k_1 = \text{coefficient of } t = np = \text{mean}, \qquad (8.5)$$

$$k_2 = \text{coefficient of } t^2/2! = np(1 - p) = npq = \text{Variance } \sigma^2$$

$\Rightarrow$

$$\text{standard deviation} = \sigma = \sqrt{(npq)}, \qquad (8.6)$$

$$k_3 = \text{coefficient of } t^3/3! = np(1 - 3p + 2p^2) = np\{1 - p - 2p(1 - p)\}$$

$$= npq(1 - 2p) = npq(q - p), \qquad (8.7)$$

$$k_4 = \text{coefficient of } t^4/4! = np(1 - p - 6p + 12p^2 - 6p^3)$$

$$= np\{q - 6p(1 - 2p + p^2)\} = np\{q - 6p(1 - p)^2\} = npq(1 - 6pq), \quad (8.8)$$

etc. Thus, from Eq. (4.2), we get the coefficient of skewness:

$$\gamma_1 = \mu_3 / \sigma^3 = k_3 / \sigma^3 = npq(q - p) / (npq)^{3/2} = (q - p) / \sqrt{(npq)}.$$

Also, from Eq. (5.1), we obtain the absolute value of kurtosis:

$$\beta_2 = \mu_4/\sigma^4 = k_4/(k_2)^2 = npq\,(1-6pq)/(npq)^2 = (1-6pq)/npq. \quad (8.9)$$

Recalling § 5, we conclude from Eq. (8.7), that the distribution is symmetric when $1-2p=0 \Rightarrow p=q=1/2$. On contrary, it is positively skew when $1-2p>0 \Rightarrow p<1/2$, i.e. $q>1/2$. In case of otherwise when $1-2p<0 \Rightarrow p>1/2$, i.e. $q<1/2$, it is negatively skew. Also, from Eqs. (5.3) and (8.9), we conclude that the distribution is leptokurtic (resp. platykurtic) when

$$1-6pq \equiv 1-6p\,(1-p) = 1-6p+p^2 > 0 \quad (\text{resp.} < 0).$$

Example 8.1. Five coins are tossed up 256 times under almost same conditions. Frequencies for '*heads*' are given as per the following distribution:

Heads:	0,	1,	2,	3,	4,	5;
Frequencies:	9,	41,	76,	93,	37,	10.

$$\left.\right\} \quad (8.10)$$

Fit a binomial distribution to draw inference about the perfectness of the coin.

Solution. Let the coins be perfect so that $p=q=1/2$. As seen above in Eq. (8.3), the frequencies in the distribution for tossing up the coin 256 times are the successive terms in the binomial expansion of

$$256.\,(1/2+1/2)^5 = (256/2^5).\,(1 + {}^5C_1 + {}^5C_2 + {}^5C_3 + {}^5C_4 + {}^5C_5)$$

$$= 8.\,(1 + 5 + 10 + 10 + 5 + 1);$$

i.e. 8, 40, 80, 80, 40 and 8 which are very close to those in Eq. (8.10). Hence, the coins seem to be perfect. //

Example 8.2. Establish the recurring relation for the binomial distribution:

$$\mu_{r+1} = pq\,\{nr\,\mu_{r-1} + d\mu_r/dp\}. \quad (8.11)$$

Solution. The frequencies f_r's of binomial distribution are given by Eq. (8.3) and the mean of the distribution is calculated vide Eq. (8.5). Hence, the moment μ_r of the distribution about the mean value given by Eq. (2.5) reads as

$$\mu_r = \sum_{k=0}^{n} {}^nC_k p^k . q^{n-k} . (k - np)^r \quad \Rightarrow$$

$$d\mu_r / dp = \sum_{k=0}^{n} {}^nC_k p^{k-1} . q^{n-k-1} . \{kq + p.(n - k).(dq/dp)\} . (k - np)^r$$

$$- nr. \sum_{k=0}^{n} {}^nC_k p^k . q^{n-k} . (k - np)^{r-1} .$$

Putting for $dq/dp = -1$ obtainable from Eq. (8.1), the terms within curly brackets in above equation simplify to $k - np$ and taking out the factor pq, the first sum on RHS simplifies to

$$(1/pq). \sum_{k=0}^{n} {}^nC_k p^k . q^{n-k} . (k - np)^{r+1} = (1/pq).\mu_{r+1},$$

while the last term becomes $- nr.\mu_{r-1}$ by Eq. (2.5). Thus, the final equation assumes the form as in Eq. (8.11). //

8.2. Poisson distribution

It is a discrete distribution relating extremely rare events but occurs frequently. Choosing n sufficiently large but p small in the binomial distribution so that the mean np is finite:

$$\lim_{n \to \infty} np = m \text{ (say), a finite (real) number.} \tag{8.12}$$

Expanding the binomial coefficients:

$${}^nC_r = n! / r! (n - r)! = n (n - 1) (n - 2) ...(n - r + 1) / r!$$

$$= n^r (1 - 1/n) (1 - 2/n) ... \{1 - (r - 1) / n\} / r!,$$

and re-writing

$$p^r.q^{n-r} = (m/n)^r.\{1 - (m/n)\}^{n-r} = (m^r/ n^r).\{1 - (m/n)\}^n .\{1 - (m/n)\}^{-r},$$

for Eqs. (8.1) and (8.12), we obtain

$$\lim_{n \to \infty} {}^nC_r .p^r.q^{n-r} = (m^r/ r!). \lim_{n \to \infty} \{1 - (m/n)\}^{n-r} = m^r.e^{-m} /r!.$$

Hence, the limit of frequency f_r for value r in a binomial distribution as given by Eq. (8.3) is

$$\lim_{n \to \infty} f_r = N.m^r.e^{-m} /r!, \tag{8.13}$$

determining the frequencies in Poisson distribution. The variable r with these frequencies is called the *Poisson variate*.

The *moment generating function* of the Poisson distribution about the origin is found by Eq. (3.1):

$$M(t) = (1/N).\sum_{r=0}^{\infty} f_r \, e^{tr} = e^{-m}.\sum_{r=0}^{\infty} (me')^r / r!$$

$$= e^{-m}.\exp(me') = \exp\{m(e'-1)\} \qquad (8.14)$$

giving cumulative generating function

$$\psi(t) = \ln M(t) = m.(e^t - 1) = m.(t + t^2/2! + \ldots + t^r/r! + \ldots). \qquad (8.15)$$

The coefficients of each term within the parentheses are m. Hence,

$$\text{Mean} = k_1 = \text{coefficient of } t = m,$$

$$\text{Variance} = k_2 = \text{coefficient of } t^2/2! = m \Rightarrow \text{s.d.} = \sigma = \sqrt{m},$$

$$k_3 = \text{coefficient of } t^3/3! = m,$$

$$k_4 = \text{coefficient of } t^4/4! = m, \text{ etc.}$$

$$(8.16)$$

$$\text{Coefficients of skewness are: } \beta_1 = (k_3)^2 / (k_2)^3 = 1/m,$$

$$\gamma_1 = \sqrt{\beta_1} = 1/\sqrt{m},$$

$$\text{Excess of kurtosis} = \gamma_2 = k_4 /(k_2)^2 = 1/m$$

$$\Rightarrow \qquad \beta_2 = \gamma_2 + 3 = 1/m + 3.$$

$$(8.17)$$

Conclusively, Poisson distribution is always skew: positively (or negatively) for corresponding values of m. It is also leptokurtic (resp. platykurtic) for positive (resp. negative) value of m.

Theorem 8.1. For a Poisson distribution the mean value and other symbols are connected by the relation

$$k_1 \sigma \gamma_1 \gamma_2 = 1. \qquad (8.18)$$

Proof. Substituting values of the symbols obtainable from Eqs. (8.16) and (8.17), one gets the result immediately. //

Example 8.3. Fit a Poisson distribution to the following data

$$\left.\begin{array}{ll} \text{Deaths:} & 0, \quad 1, \quad 2, \quad 3, \quad 4; \\[2mm] \text{Frequencies:} & 122, \ 60, \ 15, \ 2, \ 1; \end{array}\right\} \qquad (8.19)$$

and compute the theoretical frequencies.

Solution. Eq. (2.4) determines the mean:

$$m = (122.0 + 60.1 + 15.2 + 2.3 + 1.4)/(122 + 60 + 15 + 2 + 1) = 1/2 = \cdot 5$$

$$\Rightarrow \quad e^{-m} = e^{-0\cdot 5} = 1 - \cdot 5 + (\cdot 5)^2/2 - (\cdot 5)^3/6 + \ldots = 0\cdot 6065 \text{ (approx.)}$$

$$\Rightarrow \qquad N.\, e^{-m} = 200.\,(0\cdot 6065) = 121\cdot 30 \approx 121.$$

Therefore, Eq. (8.13) determines frequencies of deaths:

$$\lim_{n \to \infty} f_r = N.m^r.e^{-m}/r! = (121).\, m^r/r! \qquad (8.20)$$

approximately. Thus, we get

Frequency of zero number of deaths, (i.e. at $r = 0$) $= f_0 = 121$,

Frequency of one death $= f_1 = 121m = 121/2 = 60\cdot 5 \approx 61$,

Frequency of 2 deaths $= f_2 = 121m^2/2 = 121/8 \approx 15$,

Frequency of 3 deaths $= f_3 = 121m^3/6 = 121/48 \approx 2\cdot 52 \approx 3$,

Frequency of 4 deaths $= f_4 = 121m^4/24 = 121/384 \approx 0\cdot 31 \approx 0$.

Thus, the theoretical frequencies for $r = 0, 1, 2, 3, 4$ are 121, 61, 15, 3, 0 respectively. //

Note 8.1. For cumulative frequency 200, f_4 may also be evaluated as

$$f_4 = 200 - (f_0 + f_1 + f_2 + f_3) = 200 - (121 + 61 + 15 + 3 + 0) = 0.$$

8.3. Normal distribution

Let r be a binomial variate with mean np and variance npq, where neither p nor q is very small. The distribution

$$z = (r - np) / \sqrt{(npq)} \qquad (8.21)$$

under limiting case, when $n \to \infty$, becomes a continuous distribution called *normal distribution*. Therefore, the frequencies given by Eq. (8.3), under the change of variables affected by

$$r = np + x, \qquad (8.22)$$

where x is the deviation in the value of the variable r to mean. Making use of this change the frequency becomes

$$f_r = N \cdot {}^nC_{np+x} \cdot p^{np+x} \cdot q^{n-(np+x)}.$$

Employing the middle ordinates y_x and y_0 of histogram of the binomial variates $r = np + x$ and $r = np$ respectively for the mean, so that

$$y_x \propto N \cdot {}^nC_{np+x} \cdot p^{np+x} \cdot q^{nq-x} \quad \text{and} \quad y_0 \propto N \cdot {}^nC_{np} \cdot p^{np} \cdot q^{nq}$$

$$\Rightarrow$$

$$y_x / y_0 = \{ {}^nC_{np+x} / {}^nC_{np} \} \cdot (p/q)^x. \qquad (8.23)$$

The ratio of binomial coefficients simplifies to

$${}^nC_{np+x} / {}^nC_{np} = (np)! \cdot (nq)! / (np+x)! \cdot (nq-x)!, \qquad (8.24)$$

for Eq. (8.1). Applying Stirling's formula

$$n! = \sqrt{(2\pi)} \cdot n^{n+1/2} \cdot e^{-n},$$

in Eq. (8.24) and putting for the ratio of binomial coefficients, Eq. (8.23) reduces to

$$y_x / y_0 = \{ (np)^{np+1/2} \cdot e^{-np} \cdot (nq)^{nq+1/2} \cdot e^{-nq}$$

$$/ (np+x)^{np+x+1/2} \cdot e^{-(np+x)} \cdot (nq-x)^{nq-x+1/2} \cdot e^{-(nq-x)} \} \cdot (p/q)^x.$$

$$= 1/(1+x/np)^{np+x+1/2}. (1-x/nq)^{nq-x+1/2}$$

$$\Rightarrow$$

$$\ln (y_x/y_o) = - [(np+x+1/2).\ln (1+x/np) + (nq-x+1/2).\ln (1-x/nq)]$$

Expanding the logarithmic functions and arranging the terms in ascending powers of x, the RHS expression of above relation reduces to

$$(p-q).(x /2npq) - x^2/2npq + (x/2n)^2.(1/p^2 + 1/q^2) + o (x^3/n^2), \quad (8.25)$$

where $o (x^3/n^2)$ denotes the sum of terms involving higher powers of x. For Eqs. (8.21) and (8.22), the variable x changes into z: $x = z \sqrt{(npq)}$. Hence, the expression (8.25) changes to

$$(p-q). z / 2\sqrt{(npq)} - z^2/2 + (z^2 /4n). (q/p + p/q) + o (x^3/n^2),$$

and

$$\lim_{n \to \infty} \ln (y_z/ y_o) = - z^2/2 \Rightarrow \lim_{n \to \infty} y_z = y_o. \exp (- z^2/2).$$

Since r is a non-negative integral variate in Eq. (8.21) and the difference between two successive values of r is 1 implying the difference between two successive values of $z = 1/\sqrt{(npq)}$ as infinitesimally small for larger values of n. This makes z a continuous variate with frequency function

$$y = y_o. \exp (- z^2/2). \tag{8.26}$$

The variate z ranges from $-\infty$ to ∞ corresponding to values of $r = 0, 1, 2, 3, \ldots , \infty$.

8.4. Relative frequency function of normal distribution

Area under the frequency curve represented by Eq. (8.26) is

$$A = y_o. \int_{-\infty}^{\infty} \exp (- z^2/2). dz = 2y_o. \int_{0}^{\infty} \exp (- z^2/2). dz,$$

by a property of definite integrals, vide Eq. (1.1.7). Setting

$$z^2 = 2t \quad \Rightarrow \quad dz = dt / z = dt /\sqrt{(2t)}, \tag{8.27}$$

above area becomes

$$A = \sqrt{2}. y_o. \int_{0}^{\infty} t^{-1/2}.e^{-t}. dt = \sqrt{2}. y_o.\Gamma(1/2) = \sqrt{(2\pi)}. y_o.$$

Putting for y_o in Eq. (8.26), we get the relative frequency distribution function

$$y = \{1/\sqrt{(2\pi)}\}. \exp(-z^2/2). \tag{8.28}$$

Using properties of definite integrals, and substitutions as given by Eq. (8.27), one may evaluate the mean, variance, coefficients of skewness and kurtosis for the distribution from Eq. (3.3):

$$
\left.
\begin{aligned}
&\text{Mean} = \mu_1' = \{1/\sqrt{(2\pi)}\}. \int_{-\infty}^{\infty} z. \exp(-z^2/2). \, dz = 0, \\[6pt]
&\text{Variance} = \mu_2' = \{1/\sqrt{(2\pi)}\}. \int_{-\infty}^{\infty} z^2. \exp(-z^2/2).dz \\[6pt]
&= (2/\sqrt{\pi}). \int_{0}^{\infty} t^{1/2} e^{-t}.dt = (2/\sqrt{\pi}).\Gamma(3/2) = 1 = \sigma^2 \Rightarrow \sigma = 1, \\[6pt]
&\mu_3' = \{1/\sqrt{(2\pi)}\}. \int_{-\infty}^{\infty} z^3. \exp(-z^2/2). \, dz = 0, \\[6pt]
&\mu_4' = \{1/\sqrt{(2\pi)}\}. \int_{-\infty}^{\infty} z^4. \exp(-z^2/2).dz \\[6pt]
&= (4/\sqrt{\pi}). \int_{0}^{\infty} t^{3/2} e^{-t}.dt = (4\sqrt{\pi}). \Gamma(5/2) = 3.
\end{aligned}
\right\} \tag{8.29}
$$

Hence, Eq. (4.3) determines the coefficient of skewness

$$\gamma_1 = \mu_3'/\sigma^3 = 0, \quad \text{and} \quad \beta_2 = \mu_4'/\sigma^4 = 3, \quad \text{by Eq. (8.29).}$$

Definition 8.1. The variable in the normal distribution with zero mean and unit standard deviation is called the *standardized normal variate*.

The relative frequency for a normal variate x varying from $-\infty$ to ∞ in a normal distribution with mean m and standard deviation σ is given by

$$y = \{1/\sigma\sqrt{(2\pi)}\}. \exp\{-(x-m)^2/2\sigma^2\}. \tag{8.30}$$

Area under this curve may be seen as unity, and the transformation

$$z = (x-m)/\sigma$$

transforms the normal variate x into standardized normal variate z. Also, assuming the mean m at origin, Eq. (8.30) changes to the normal distribution given by Eq. (5.2).

The graph of such a frequency curve is a *normal curve* shown in the Fig. 8.1. Areas under this curve included between $-\sigma \le x \le \sigma$, $-2\sigma \le x \le 2\sigma$, and $-3\sigma \le x \le 3\sigma$ are approximately 68%, 95% and 99% of the whole area

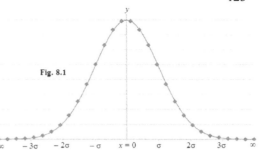

Fig. 8.1

under the curve. Moment generating function of the distribution is given by Eq. (3.1):

$$M(t) = \{1 / \sigma \sqrt{(2\pi)}\} . \int_{-\infty}^{\infty} e^{tx} . \exp(-x^2/2\sigma^2) . dx. \qquad (8.31)$$

Combining the powers of e and reducing them in a perfect square form:

$$-x^2/2\sigma^2 + t x = -\{x^2 - 2(t\sigma^2)x + t^2 . \sigma^4\}/2\sigma^2 + (t\sigma)^2/2$$

$$= -(x - t\sigma^2)^2/2\sigma^2 + (t\sigma)^2/2 = -u + (t\sigma)^2/2,$$

where

$$(x - t\sigma^2)^2 = 2\sigma^2 . u \quad \Rightarrow \quad (x - t\sigma^2) . dx = \sigma^2 . du$$

$$\Rightarrow$$

$$dx = \sigma^2 . du / (x - t\sigma^2) = \sigma . du / \sqrt{(2u)} = (\sigma/\sqrt{2}) . u^{-1/2} . du,$$

Eq. (8.31) simplifies to

$$M(t) = (1/\sqrt{\pi}) . \exp(t^2 \sigma^2/2) . \int_0^{\infty} u^{-1/2} . e^{-u} . du$$

$$= (1/\sqrt{\pi}) . \exp(t^2 \sigma^2/2) . \Gamma(1/2) = \exp(t^2 \sigma^2/2).$$

Hence, Eq. (8.15) determines the cumulative generating function:

$$\psi(t) = \ln M(t) = t^2 \sigma^2/2. \qquad (8.32)$$

This determines the mean, variance etc. of the distribution:

Mean $= k_1 =$ coefficient of $t = 0$,

Variance $= k_2 =$ coefficient of $t^2/2! = \sigma^2$, $\qquad (8.33)$

$k_3 =$ coefficient of $t^3/3! = 0$, $k_4 =$ coefficient of $t^4/4! = 0$, etc.

A standard normal distribution with mean 0 and variance 1 is denoted by $N(0, 1)$ while a general normal distribution with mean m and variance σ is denoted by $N(m, \sigma)$. The m.g.f. for these distributions are $\exp(t^2/2)$ and $\exp(m t + t^2 \sigma^2/2)$ respectively.

Example 8.4. Show that the mean deviation from the mean of the normal distribution is about $4/5^{th}$ part of its standard deviation.

Solution. Let us consider a normal distribution given by Eq. (5.2). As seen in Eq. (8.33), the mean of the distribution is zero. Hence, the mean deviation from the mean is

$$\{1/\sigma\,\sqrt{(2\pi)}\}.\int_{-\infty}^{\infty} |x|.\exp(-x^2/2\sigma^2).\,dx. \tag{8.34}$$

Breaking the range of integration and using properties of mod function, the integral reduces to

$$-\int_{-\infty}^{0} x.\exp(-x^2/2\sigma^2).\,dx + \int_{0}^{\infty} x.\exp(-x^2/2\sigma^2).\,dx.$$

Setting $-x = v \Rightarrow -dx = dv$, and using the properties of definite integrals, the first integral equals to the second one. Further, putting

$$x^2 = 2\sigma^2 u \qquad \Rightarrow \qquad x.\,dx = \sigma^2.\,du,$$

$$\{2\sigma/\sqrt{(2\pi)}\}.\int_{0}^{\infty} e^{-u}.\,du = \sigma\sqrt{(2/\pi)}.\left[-e^{-u}\right]_{0}^{\infty} = \sigma\,\sqrt{(2/\pi)} \approx (4/5)\,\sigma.\;//$$

Example 8.5. In a normal distribution of population given by

$$f(t) = \{1/\sqrt{(2\pi)}\}.\int_{0}^{t} \exp(-t^2/2).\,dt, \tag{8.35}$$

together with $f(0·5) = 0·19$ and $f(1·40) = 0·42$, 31% persons are below 45 years while 8% are above 64. Find the mean and standard deviation.

Solution. Eq. (8.28) represents a normal curve for which mean $= 0$, and standard deviation $= 1$ (cf. Eq. (8.29)). The function $f(t)$ in Eq. (8.35) represents the total area under above normal curve when $t \to \infty$. Applying the substitution as in Eq. (8.27) this area becomes

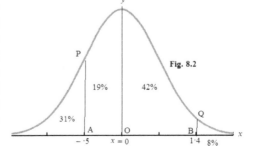

Fig. 8.2

$$(1/\sqrt{\pi}).\int_0^\infty u^{-1/2}.e^{-u}.du = (1/\sqrt{\pi}).\Gamma(1/2) = 1. \qquad (8.36)$$

Let AP and BQ represent the ordinates of the curve so that area under the curve lying to the left of AP is 31%, i.e. 0·31 units, and that to the right of BQ is 8%, i.e. 0·08. Hence, the area between AP and y-axis is 50 − 31 = 19%, i.e. 0·19 and that between y-axis and BQ is 50 − 8 = 42%, i.e. 0·42.

Let m be the mean and σ the standard deviation of the normal distribution under the question. Comparing the present curve (Fig. 8.2) with the normal curve represented by Eq. (8.28), i.e.

$$y = \{1/\sqrt{(2\pi)}\}.\exp(-x^2/2), \qquad (8.28a)$$

in terms of the variable x, we have

$(45 - m)/\sigma$ = the value of x measuring 0·19 area, i.e. − 0·5

$\Rightarrow$

$$45 - m = -0.5\,\sigma,$$

and

$(64 - m)/\sigma$ = the value of x measuring 0·42 area, i.e. 1·4

$\Rightarrow$

$$64 - m = 1.4\,\sigma.$$

Solving these linear equations for m and σ, we get $m = 50$ and $\sigma = 10$. //

CHAPTER 10

TESTS OF SIGNIFICANCE

§ 1. Introduction

When the population is large its complete enumeration is not possible then sampling is preferred, which is of four kinds:

(i) Purposive - with a specific purpose in view,

(ii) Random - wherein each unit of population has an equal chance for its inclusion,

(iii) Stratified – wherein entire heterogeneous population is divided into several homogeneous groups like strata of diverse nature. The units are sampled at random from each stratum.

(iv) Systematic: It is a statistical method involving the selection of elements from an ordered sampling frame. The most common form of systematic sampling is an equi-probability method.

The parameters, i.e. the constants of the population: mean $\bar{x}$, variance σ^2, are said *statistic*. The set of values of the statistic constitute the sampling distribution of the statistic. The standard deviation of the sampling distribution of a statistic is known as *standard error*. As such, the study of 'tests of significance' is an important aspect of sampling theory. For large n, almost all the distributions (binomial, Poisson, etc.) being close to normal distribution, 'normal test of significance' is used for large samples. Some of the well-known tests of significance for study of such differences, for instance, *chi-square* and t-test are discussed here.

1.1. Null hypothesis

To apply the tests of significance we first set up a hypothesis, i.e. a definite statement about the population parameters. This is usually a hypothesis of no difference and is called *null hypothesis*. It is denoted by H_o and is tested for possible rejection under the assumption that it is true. Any other hypothesis complimentary to H_o is an alternative hypothesis and is denoted by H_1. For example, to test H_o, that is to say if population has mean m_o, alternative hypotheses could be that it has:

(i) mean $\neq m_0$, i.e. mean $> m_0$, or mean $< m_0$;

(ii) mean $> m_0$, (iii) mean $< m_0$.

Symbolically, the hypotheses are expressed in the following ways:

Null hypothesis: $H_0 : m = m_0$;

Alternative hypotheses: $H_1 : m \neq m_0$; $H_1 : m > m_0$; $H_1 : m < m_0$.

Note 1.1. Alternative hypothesis decides the shape of the distribution curve whether it has a single tail or alike.

1.2. Pearson's chi-square statistic

In the end of 19th century, the German mathematician Karl Pearson noticed the existence of significant *skewness* within some biological observations. To model the observations whether *normal* or *skewed*, Pearson, introduced a family of continuous probability distributions (both normal and skewed) in a series of articles published from 1893 to 1916. He proposed a method of statistical analysis to model the observation and performing the test of goodness of fit to determine how well the model and the observation really fit. This statistic was called after him as Pearson's chi-square statistic or in short as chi-square statistic.

Chi-square (χ^2) statistic is a test that measures how expectations compare to actual observed data. The square of a standard normal variate is known as a chi-square variate with one degree of freedom.

Definition 1.1. Excess of number of observations over the number of independent constraints is called the degree of freedom of the variate.

Let us consider n independent normal variates with mean m_i and variance $(\sigma_i)^2$. The following variate

$$\chi^2 = \sum_{i=1}^{n}\{(x_i - m_i)/\sigma_i\}^2 = \sum_{i=1}^{n} u_i^2, \qquad (1.1)$$

where

$$u_i = (x_i - m_i)/\sigma_i \quad \Rightarrow \quad du_i = dx_i/\sigma_i. \qquad (1.2)$$

function of the distribution about the origin may be found by Eq. (6.3.1):

$$M(\chi^2, t) = \int_{-\infty}^{\infty} \exp(t \, u_i^2). f(x_i) \, dx_i, \qquad (1.3)$$

$$= \{1/\sigma_i \sqrt{(2\pi)}\}. \int_{-\infty}^{\infty} \exp(t \, u_i^2). \exp\{-(x_i - m_i)^2 / 2(\sigma_i)^2\}. dx_i$$

$$= \{1/\sqrt{(2\pi)}\}. \int_{-\infty}^{\infty} \exp\{-(1 - 2t) \, u_i^2 / 2\}. du_i,$$

by Eqs. (9.8.30) and (1.2). Using properties of definite integrals vide Eq. (1.1.7) and putting

$$(1 - 2t) \, u_i^2 / 2 = v \qquad \Rightarrow \qquad du_i = dv / \sqrt{\{2(1 - 2t). v\}},$$

above integral reduces to

$$M(\chi^2, t) = \{1/\sqrt{\pi}(1 - 2t)\}. \int_0^{\infty} v^{-1/2}. e^{-v}. dv = (1 - 2t)^{-1/2},$$

by Eq. (9.8.36). Notably, this is also the m.g.f. of a Gamma variate with parameters $1/2$ and $n/2$. Hence, we have the

Theorem. 1.1. The chi-square (χ^2) is a Gamma variate.

§ 2. Conditions for validity of chi-square test

(1) Being an approximate test for larger values of n there must hold the following conditions:

(i) The sample observations should be independent;

(ii) Constraints on the cell frequencies should be linear;

(iii) Total frequency (N) should be large enough, say over 50;

(iv) No theoretical cell frequency should be lower than 5 as χ^2- distribution is essentially continuous but it cannot maintain continuity in case cell frequency lesser than 5.

(2) Also, χ^2- test depends only on the set of observed and expected frequencies and on the degree of freedom. It does not make any assumption regarding the parent population providing the observations.

(3) The critical values of χ^2 increase as per increasing degree of freedom n but level of significance decreases.

2.1. Applications of chi-square distribution

It has wide applications in Statistics of which the following ones are noteworthy:

(i) To test if hypothetical value of population variance is σ_0^2;

(ii) To test the goodness of fit;

(iii) To test the independence of attributes;

(iv) To test the homogeneity of independent estimates of population variance and of correlation coeffcient;

(v) To combine various possibilities obtained from independent experiments to give a single test of significance.

Example 2.1. The precision of an instrument is believably not more than 0·16. Write down the null and alternative hypotheses to test the belief. Giving 11 trials on the instrument carry out the test at 1% level:

$$2·5, \ 2·3, \ 2·4, \ 2·3, \ 2·5, \ 2·7, \ 2·5, \ 2·6, \ 2·6, \ 2·7, \ 2·5.$$

(Note that precision is measured by the variance σ^2 of observation from the mean value $\bar{x}$.)

Solution. The null hypothesis (H_0) is: $\sigma^2 = 0·16$, and the alternative hypothesis (H_1) is: $\sigma^2 > 0·16$. The (arithmetic) mean of above 11 observations, by Eq. (9.2.4), is $\bar{x} = 27·6/11 = 2·51$. Sample variance (at 1%) from the mean value is computed as per the following table:

x_i	$x_i - \bar{x}$	$(x_i - \bar{x})^2$
2·5	$2·5 - 2·51 = -0·01$	0·0001
2·3	$2·3 - 2·51 = -0·21$	0·0441
2·4	$2·4 - 2·51 = -0·11$	0·0121
2·3	$2·3 - 2·51 = -0·21$	0·0441

2·5	2·5– 2·51 = – 0·01	0·0001
2·7	2·7 – 2·51 = 0·19	0·0361
2·5	2·5– 2·51 = – 0·01	0·0001
2·6	2·6 – 2·51 = 0·09	0·0081
2·6	2·6 – 2·51 = 0·09	0·0081
2·7	2·7 – 2·51 = 0·19	0·0361
2·5	2·5– 2·51 = – 0·01	0·0001

Therefore, $\sum (x_i - \bar{x})^2 = 0.1891$ for which, under null hypothesis, Eq. (1.1) determines

$$\chi^2 = \sum_{i=1}^{11} \{(x_i - m_i)/\sigma_i\}^2 = 0.1891/0.16 = 1.182. \tag{2.1}$$

This follows chi-square distribution with degree of freedom = 10 (cf. Defn. 1.1). Since the computed value of χ^2 in Eq. (2.1) is lesser than the tabulated value 2·5 for degree of freedom = 10 at 1% level of significance. Therefore, H_0 may be accepted to conclude the consistency of data with the hypothesis and the precision of the instrument is 0·16. //

Example 2.2. Distribution of digits in numbers chosen at random from a telephone directory have the following frequencies (f_i):

dig-its	0	1	2	3	4	5	6	7	8	9
f_i	1026	1107	997	966	1075	933	1107	972	964	853

Test if digits occur equally frequently in the directory? Given 16·919 as the tabulated value of χ^2 at 5% (i.e. 0·05) level of significance for 9 degree of freedom.

Solution. Assuming that the digits occur equally frequently, i.e. accepting the null hypothesis H_0. As such, the expected frequency for each digit = sum of frequencies / number of digits = 1,000. We compute the data as per following table:

Digits	Frequency		$E = (f_o - f_e)^2$	E / f_e
	Observed f_o	Expected f_e		
0	1,026	1,000	676	0·676
1	1,107	1,000	11,449	11·449
2	997	1,000	9	0·009
3	966	1,000	1,156	1·156
4	1,075	1,000	5,625	5·625
5	933	1,000	4,489	4·489
6	1,107	1,000	11,449	11·449
7	972	1,000	784	0·784
8	964	1,000	1.296	1·296
9	853	1,000	21,609	21·609
Total		$\sum f_e = 10,000$		58·542

Therefore,

$$\chi^2 = \sum (f_o - f_e)^2 / f_e = 58 \cdot 542. \tag{2.2}$$

The single linear constraint is $\sum f_e = 10,000$ so, by Defn. 1.1, degree of freedom = $10 - 1 = 9$. As given, the tabulated value of χ^2 (0·05) for 9 degree of freedom is 16·919. Thus, the computed value of χ^2 in Eq. (2.2) at this degree of freedom is much higher than the given value and so it is highly significant. That leads the rejection of null hypothesis. Hence, digits are not uniformly distributed in the directory. //

Example 2.3. Number of air accidents occurring during a certain week are given by the following table:

Days	Mon	Tues	Wed	Thurs	Fri	Sat
No. of accicents	14	18	12	11	15	14

Check if the accidents are uniformly distributed over the week? Values of χ^2 significant at degree of freedom: 5, 6, 7 are 11·07, 12·59 and 14·07 respectively at 5% level of significance.

Solution. Assuming that the accidents are uniformly distributed over week, i.e. accepting the null hypothesis H_0. As such, the expected frequency of accidents each day = sum of frequencies / number of days = 84/6 =14. We compute the data as per following table:

Days	Frequency of accidents		$E = (f_o - f_e)^2$	E / f_e
	Observed f_o	Expected f_e		
Mon.	14	14	0	0·000
Tues.	18	14	16	1·143
Wed.	12	14	4	0·286
Thurs.	11	14	9	0·643
Friday	15	14	1	0·071
Sat.	14	14	0	0·000
Total		$\sum f_e = 84$		2·143

Therefore,

$$\chi^2 = \sum (f_o - f_e)^2 / f_e = 2 \cdot 143. \tag{2.3}$$

There is a single linear constraint $\sum f_e = 84$ so, By Defn. 1.1, the degree of freedom for 6 days' observations = 6 − 1 = 5. Also, the tabulated value of χ^2 at 5% (= 0·05) level of significance for 5 degree of freedom is given 11·07. Thus, the computed value of χ^2 in Eq. (2.3) is much lower than the given value so it is highly insignificant. Thus, null hypothesis is acceptable leading to the fact that accidents are uniformly distributed over the week. //

§ 3. Student's *t*-test

The *t*-test is a statistical hypothesis test where the test statistic follows a Student's *t*-distribution under the null hypothesis. It was introduced by a British chemist (William Sealy Gosset) working for the Guinness brewery in Dublin in 1908. He used *Student* as his pen name. Gosset devised the *t*-test as an economical way to monitor the quality of stout. He published his research work in the journal *Biometrika* in 1908. The Company did not permit its employees to publish their findings in their names, so Gosset published his statistical work under the pseudonym *Student*, which is not to be confused with the literary word student.

Let x_i, $i = 1, 2, \ldots, n$ be a random sample of size n from a normal population with mean m and variance σ^2. In one-sample t-test

$$t = (\bar{x} - m) / (\sigma / \sqrt{n}), \tag{3.1}$$

where $\bar{x}$, as per Eq. (9.2.4), is the sample mean:

$$\bar{x} = (1/n).\sum_{i=1}^{n} x_i, \quad \text{and} \quad s^2 = \sum_{i=1}^{n}(x_i - \bar{x})^2/(n-1), \tag{3.2}$$

is an unbiased estimate of the population variance σ^2. It follows Student's t-distribution with $\upsilon = n - 1$ degree of freedom and probability density function

$$f(t) = 1/\sqrt{\upsilon}.B(1/2, \upsilon/2).(1 + t^2/\upsilon)^{(\upsilon+1)/2}, \quad -\infty < t < \infty. \tag{3.3}$$

Particulary, for $\upsilon = 1$, i.e. $n = 2$, above function reduces to

$$f(t) = 1/B(1/2, 1/2).(1 + t^2) = (1/\pi)(1 + t^2)^{-1/2}, \quad -\infty < t < \infty, \tag{3.4}$$

by a property of beta function:

$$B(1/2, 1/2) = \Gamma(1/2).\Gamma(1/2)/\Gamma(1/2 + 1/2) = \sqrt{\pi}.\sqrt{\pi}/\Gamma(1) = \pi. \tag{3.5}$$

It may be noted that above value of $f(t)$ is the probability density function of standard Cauchy distribution. Thus, for $\upsilon = 1$, Student's t-distribution reduces to Cauchy distribution.

Note 3.1. A statistic t following Student's t-distribution with n degree of freedom will be abbreviated as $t \sim t_n$.

3.1. Assumptions for Student's t-test

(i)   The parent population providing the sample is normal;
(ii) The sample is drawn at random;
(iii) The standard deviation σ of the population is unknown.

Example 3.1. The average weekly sale of soap bars in each Departmental store is 146·3 bars. After an advertising campaign, the average weekly sale in 22 stores in a particular week increased to 153·7 with a standard deviation 17·2. Was the advertising campaign successful?

Solution. Given that $n = 22$, $\bar{x} = 153\cdot7$, $s = 17\cdot2$. Accepting the null hypothesis H_o that advertising is not successful, so that mean $m = 146\cdot3$, an alternative hypothesis is $H_1 : m > 146.3$. Under the null hypothesis

$$t = (\bar{x} - m) / (s /\sqrt{n}) = (153\cdot7 - 146\cdot3).(\sqrt{21}) / 17\cdot2 = 1\cdot97. \quad (3.6)$$

The tabulated value of t for 21 degree of freedom at 5% level of significance is $1\cdot72$, but computed value vide Eq. (3.6), is larger, it is significant. Hence, null hypothesis is not acceptable leading to the success of advertising campaign. //

3.2. Applications of *t*-distribution

It has wide applications in Statistics of which few are detailed below:

(i)   To test sample mean, $\bar{x}$ differs significantly from the hypothetical value m of the population mean;

(ii) To test the significant difference between two sample means;

(iii) To test the significance of an observed sample correlation coefficient and sample regression coefficient;

(iv) To test the significance of observed partial and multiple correlation coefficients.

Example 3.2. The heights of 10 boys are 70, 67, 62, 68, 61, 68, 70, 64, 64, 66 cms. Is it reasonable to take average height greater than 64 cms.? Test at 5% significance level assuming that for 9 degrees of freedom the tabulated value of t is $1\cdot83$.

Solution. Accepting the null hypothesis H_o that the average (i.e. mean) height $m = 64$ cms., an alternative hypothesis is $H_1 : m > 64$. By Eq. (3.2), we get the sample mean

$$\bar{x} = \sum_{i=1}^{10} x_i / 10$$

$$= (70 + 67 + 62 + 68 + 61 + 68 + 70 + 64 + 64 + 66)/10 = 66,$$

and compute the values of $x - \bar{x}$ as per the following table:

x	70	67	62	68	61	68	70	64	64	66
$x - \bar{x}$	4	1	-4	2	-5	2	4	-2	-2	0
$(x - \bar{x})^2$	16	1	16	4	25	4	16	4	4	0

Hence, again by Eq. (3.2), there follows

$$s^2 = \sum_{i=1}^{10}(x_i - \bar{x})^2/(n-1) = 90/9 = 10.$$

Under the null hypothesis H_0 the test statistic, by Eq. (3.1), is

$$t = (\bar{x} - m)/(s/\sqrt{n}) = (66 - 64)/(\sqrt{10}/\sqrt{10}) = 2,$$

which is larger than the given tabulated value of t for 9 degree of free-dom at 5% level of significance, i.e. 1·83. So, the testing hypothesis is significant and is not acceptable leading to the alternative hypothesis. //

§ 4. Fisher's *t*-test

It is the ratio of a standard normal variate to the square-root of an independent chi-square variate (divided by its degree of freedom):

$$t = \xi/\sqrt{(\chi^2/n)}, \tag{4.1}$$

where ξ is a standard normal variate and χ^2 is an independent chi-square variate with degree of freedom n.

Note 4.1. Degree of freedom of χ^2 variate also acts as degree of free-dom in Fisher's *t*-distribution.

Note 4.2. Student's *t*-distribution may be regarded as a special case of Fisher's *t*-distribution for

$$\bar{x} \sim N(m, \sigma^2/n), \quad \xi = (\bar{x} - m)/\sqrt{(\sigma^2/n)} \sim N(0, 1), \tag{4.2}$$

and $\quad \chi^2 = ns^2/\sigma^2 = \sum_{i=1}^{n}(x_i - \bar{x})^2/\sigma^2, \quad$ by Eq. (3.2), $\tag{4.3}$

is an independent chi-square variate with $n - 1$ degree of freedom.

BIBLIOGRAPHY

1. Airlinghaus, Sandra: *Practical Handbook of Curve Fitting*, CRC Press, U.S.A., 1994.

2. Atkinson, Kendall E.: *An Introduction to Numerical Analysis*, John Wiley and Sons, New York (U.S.A.), 1978.

3. Beerends, R. J.; Ter Morsche, H. G. and Van de Vrie, E. M.: *Fourier and Laplace Transforms*, Cambridge University Press, Cambridge (U.K.), 2003.

4. Chapra, S.C and Raymond P. Canale: *Numerical methods for Engineers*, Tata McGraw-Hill, New Delhi (India), 5th ed., 2006.

5. Dantzig, George B.: *Linear Programming and Extensions*, Princeton University Press, New Jersey (U.S.A.), 1998.

6. Grewal, B.S.: *Higher Engineering Mathematics*, Khanna Publishers, Delhi, (India) 34th ed., 1998.

7. Izenman, Alan J.: *Modern Multivariate Statistical Techniques*, Springer, 2008.

8. Jain, M.K.; Iyengar, S.R.K. and Jain, R.K.: *Numerical Methods for Engineers and Scientists*, Wiley Eastern Ltd., New Delhi (India), 1985.

9. Kresyzig, Erwin: *Advanced Engineering Mathematics*, John Wiley & Sons, INC, New York (USA), 7th ed., 1993.

10. Kumar, Sanjeev and Verma, V.S.: *Computer Based Numerical and Statistical Techniques*, Ram Prasad & Sons, Agra (India), 4th ed., 200. , ISBN 8189352-07-5.

11. Misra, R.B.: *Analytical Geometry of Planes and Solids*, Hardwari Publications, Allahabad (India), 2004, pp. xiv +500, ISBN 81-

88574-01-5.

12. Misra, R.B.: *A First Course on Calculus with Applications to Differential Equations*, Lambert Academic Publishers, Saarbrücken (Germany), 2010, ISBN 978-3-8433-7871-0.

13. Misra, R. B.: *Laplace Transform, Differential Equations and Fourier Series*, ibid, 2010, ISBN 978-3-8433-8328-8.

14. Misra, R.B.: *Numerical Analysis for Solution of Ordinary Differential Equations*, ibid, 2010, ISBN 978-3-8433-8489-6.

15. Misra, R.B.: *Advanced Integral Calculus*, ibid, 2011, ISBN 978-3-8443-1916-3.

16. Misra, R.B.: *Advanced Applied Mathematics*, Central West Pulishing, Orange, NSW, Australia, 2018, ISBN 978-1-925823-11-0.

17. Misra, R.B.: *Mathematics for Engineers & Physicists – Pt.* 1, ibid, 2019, pp. xiv + 306, ISBN (print): 978-1-925823-51-6, ISBN (e-book): 978-1-925823-50-9.

18. Misra, R.B.: *Mathematics for Engineers & Physicists – Pt.* 2, ibid, 2019, pp. xiv + 326, ISBN (print): 978-1-925823-53-0, ISBN (e-book): 978-1-925823-52-3.

19. Sastry, S.S.: *Introductory Methods of Numerical Analysis*, Prentice-Hall of India Pvt. Ltd., New Delhi (India), 4th ed., 2005, ISBN 81-203-2761-6.

20. Scheid, Francis: *Theory and Problems of Numerical Analysis*, Schaum Series, McGraw-Hill, New York (U.S.A.), 1968.

21. Spivak, Michael: *Calculus*, Publish or Perish, Inc. Houston (U.S.A.), 3rd ed., 1994 (updated, 2008).

22. Stroud, K.A.: *Further Engineering Mathematics*, Macmillan Publishers, Ltd., London (U.K.), 2nd ed., 1990.

23. Taboga, Marco: *Lectures on Probability Theory and Mathematical Statistics*, Create Space Independent Publishing Platform, 3rd ed., 2017, ISBN: 978-19813691-95.

24. Taylor, John K. and Cihon, Cheryl: *Statistical Techniques for Data Analysis*, Chapman & Hall, CRC Press, Boca Raton, Fl. (U.S.A.), 2nd ed., 2004, ISBN: 978-1584883852.

25. Vanderbei, Robert J.: *Linear programming: Foundations and Extensions*, Kluwer Academic Publishers, Dordrecht (Holland), 2nd ed., 2001.

INDEX